Be a WINNER!

A Science Teacher's Guide to Writing Successful Grant Proposals

Be a WINNER!

A Science Teacher's Guide to Writing Successful Grant Proposals

Patty McGinnis
Kitchka Petrova

NSTApress
National Science Teachers Association
Arlington, Virginia

Claire Reinburg, Director
Wendy Rubin, Managing Editor
Rachel Ledbetter, Associate Editor
Amanda O'Brien, Associate Editor
Donna Yudkin, Book Acquisitions Coordinator

ART AND DESIGN
Will Thomas Jr., Director

PRINTING AND PRODUCTION
Catherine Lorrain, Director

NATIONAL SCIENCE TEACHERS ASSOCIATION
David L. Evans, Executive Director
David Beacom, Publisher

1840 Wilson Blvd., Arlington, VA 22201
www.nsta.org/store
For customer service inquiries, please call 800-277-5300.

Catologing-in-Publication Data are available from the Library of Congress.
Names: McGinnis, Patty, 1958- | Petrova, Kitchka, 1958-
Title: Be a winner! : a science teacher's guide to writing successful grant
    proposals / Patty McGinnis and Kitchka Petrova.
Other titles: Writing successful grant proposals
Description: Arlington, VA : National Science Teachers Association, [2016] |
    Includes bibliographical references and index. | Description based on
    print version record and CIP data provided by publisher; resource not
    viewed.
Identifiers: LCCN 2016018353 (print) | LCCN 2016008716 (ebook) | ISBN
    9781681400020 (epub) | ISBN 9781681400013 (print) | ISBN 9781681400020
    (e-book)
Subjects: LCSH: Proposal writing in research. | Science--Research grants.
Classification: LCC Q180.55.P7 (print) | LCC Q180.55.P7 M34 2016 (ebook) |
    DDC 507.1/2--dc23
LC record available at https://lccn.loc.gov/2016018353

Dedication

To our students, who were the inspiration for this book.

CONTENTS

CONTENTS

PREFACE

These are truly exciting times for science educators. Emerging technologies, the *Common Core State Standards*, and the *Next Generation Science Standards* are affecting the delivery of science instruction in this country and have combined in a perfect storm of opportunity for you and your students. No matter what you aspire to accomplish, whether it is to acquire the funds to conduct innovative investigations in your classroom, enhance your professional skills, or create a new informal learning opportunity for your students, this book is your guide to successful grant proposal writing.

As frequent workshop presenters and authors of a popular short course about writing grant proposals, we realize that many teachers find the process intimidating. Over the last two decades, we have accumulated a great deal of experience writing grant proposals and receiving grants for projects in our classrooms. We are familiar with what a sound K–12 classroom grant proposal looks like, having served on grant review committees over the years and having witnessed firsthand the enthusiasm that new equipment or otherwise financially impossible opportunities can bring to our students. We have brought thousands of dollars into our respective schools, have received funding to travel to foreign countries for the purpose of working on educational and research projects, and have been selected for participation in numerous competitive professional development opportunities that required us to clearly convey our passion for improving student learning. All of these activities are within your reach, and this book will help you achieve them.

Although there are other books related to the topic of grant proposal writing, this book is unique in that it addresses the specifics of acquiring grants for the purpose of incorporating innovative science, technology, engineering, and mathematics (STEM) experiences in K–12 classrooms. We strongly believe classroom science teachers, science coaches, district curriculum developers, and informal science educators will find this book valuable and useful for increasing the educational opportunities of students in the science classroom.

ACKNOWLEDGMENTS

The completion of this book was made possible through the efforts of several individuals. We would like to thank the anonymous reviewers who provided feedback on our initial book proposal and the completed manuscript. Their comments and suggestions helped us examine our ideas and writing from different perspectives and ultimately contributed to the creation of a book designed to motivate and guide K–12 teachers of science through the grant proposal writing process.

We would also like to express our sincere appreciation to National Science Teachers Association (NSTA) editors Claire Reinburg and Wendy Rubin for answering our questions throughout the writing process and to Amanda O'Brien for carrying out the final editing. We would also like to thank the copyeditor, Teresa Barensfeld. A special thank you to Janeen Marzewski for reading and providing suggestions on our manuscript. Finally, thank you to our families for their support and encouragement throughout the writing process.

ABOUT THE AUTHORS

Patty McGinnis is a National Board Certified Teacher with more than 25 years of teaching experience at the grade 7–12 level. She has recieved numerous grants and awards that have engaged her students in innovative science investigations. Patty teaches at Arcola Intermediate School in Eagleville, Pennsylvania, and has an EdD in educational technology from Boise State University. Her interests include the use of technology in supporting science practices. Patty is a frequent presenter at science conferences and served as National Science Teachers Association's division director for middle-level science teaching from 2012 to 2015. She is thankful for the support of her husband, Bob, and of her three incredible children, Kathleen, Matthew, and Marybeth.

Kitchka Petrova is a National Board Certified Teacher in early adolescence science. She holds a PhD in microbiology from Moscow University M. V. Lomonossov, Russia, and worked as a research scientist in Bulgaria prior to immigrating to the United States. She was a middle school science teacher in private and public schools in Miami-Dade County, Florida, for 12 years. During that time, she received funding from local, state, and national organizations to support project-based inquiry learning for her students. During the 2008–2009 school year, she served as an Albert Einstein Distinguished Educator Fellow at the National Science Foundation in Arlington, Virginia. Kitchka is currently pursuing a doctoral degree in education policy and evaluation at Florida State University, with a focus on science, technology, engineering, and mathematics (STEM) education policies and initiatives.

INTRODUCTION

Congratulations! By opening this book you are well on your way to making your professional dreams come true! As a science educator, you are concerned with the state of science education in our K–12 schools, and you understand the importance of facilitating your students' science learning through the science and engineering practices identified in the *Next Generation Science Standards* (*NGSS*). Unfortunately, funds for purchasing materials are not always available in schools, thus requiring you to seek outside funding opportunities. Given the economic situation of many school districts, it is more imperative than ever to master the art of grant proposal writing to secure funds for innovative classroom projects. Although intimidating, obtaining a grant to carry out your dream *is* within your scope. This book is designed to guide you through the process of writing a successful grant proposal and encourage you to apply your newly developed skills to pursue other professional development opportunities.

This book is aimed specifically toward the K–12 science educators who are interested in obtaining classroom grants for the purpose of extending the learning opportunities for their students and themselves. This might be achieved by adding an expensive piece of equipment to the science lab, taking students on a research field trip, obtaining funding to attend a professional development activity or event that will enhance your teaching, or seeking an opportunity available specifically to science educators.

Consider the following points:

- Many grant programs do not receive enough qualified proposals.
- If you don't apply, you'll never win!
- A grant may re-energize your teaching.
- The more you write, the easier it gets.

Our goal in writing this book was to take the mystery out of the grant proposal writing process by helping you learn how to address typical grant proposal components in your writing. You'll also glean tips regarding how to locate funding opportunities at the local, state, and national level and learn how to tailor your idea to a funding agency's requirements. Although we can't guarantee that you'll enjoy the writing process (writing is hard work!), we are certain you will be proud of your end product. We use a workbook approach that explains how to write the components typically required by funders of K–12 classroom grants. Chapter 1 outlines the reasons for writing grant proposals and seeking funding. The outcomes for students and educators are highlighted as a way to motivate and inspire you to pursue funding for your innovative teaching ideas. Chapter 2 describes what is meant by a *grant* and will help you to identify funding organizations and sources of inspiration for potential projects. In Chapter 3, you'll learn why it is important to align your proposal with the funding agency requirements, and Chapter 4 walks you through developing and writing the standard grant components. Chapter 5 addresses supplemental grant components that funding agencies may require, such as lesson plans, letters of support, and project vitae. The use of the *NGSS* as a template for your grant proposal is addressed in Chapter 6. Chapter 7 provides tips concerning proposal submission along with how to deal with rejection from funding agencies. Chapter 8 contains advice for implementing your funded project; topics such as how to work with your school district to set up a designated account for your grant funds and how to deal with unexpected or unforeseen challenges are addressed. You'll find that the skills required for successful grant proposal writing can be applied to numerous other opportunities for K–12 teachers. A variety of professional opportunities, such as research experiences, fellowships, and National Science Teachers Association (NSTA) recognition awards, are highlighted in Chapter 9. The closing chapter offers some final words of advice regarding how to pursue these opportunities and how to work collaboratively with colleagues, administrators, and parents.

So what's stopping you? There is no better time than now to make your dreams come true!

The Top 10 Reasons to Write a Grant Proposal

CHAPTER 1

You may have a multitude of reasons for seeking a grant, and no doubt you already have an idea for a grant proposal in mind. Although the primary goal for most science teachers is to improve student learning, there are many other reasons for writing a grant, including some that you may not have previously considered. As you read over the following list, open your mind to the possibilities that a grant can have to help you positively influence your students' growth while expanding your professional life. You'll no doubt begin visualizing projects, investigations, and opportunities that extend well beyond your initial ideas.

Reason 1: To Nurture and Inspire Your Students' Appreciation of Science

Whether you are a member of a teaching team or an individual teacher, your primary focus is to increase learning within your classroom. A grant can make it possible for you to buy supplies, equipment, and other materials that will enrich your classroom instruction and improve student learning. As a teacher of science, you have a tremendous advantage over other subject-area teachers. Science is all around us. It pervades our everyday lives, making it relatively easy to connect scientific principles to real-world experiences—something that many funding agencies look for when evaluating a grant proposal. Such agencies are extremely supportive of projects that promote student understanding and learning of science through authentic experiences and will gladly provide funds for the purchase of necessary equipment and supplies. Additionally, it is relatively easy to align either the *Next Generation Science Standards* (*NGSS*) or your state science standards to a funding agency's mission when submitting a grant proposal.

Reason 2: To Engage Students in Science and Engineering Practices

A good example of a classroom investigation that involved a real-world connection to physical science is the funded *Let's Go Solar* proposal discussed in Chapter 6 (p. 65). Using grant-funded solar houses, students were able to observe the greenhouse effect and learn about alternative energy resources while deepening their understanding of energy and the process of energy transfer. The grant also paid for solar cars, which were used to measure speed and velocity in student-designed experiments that investigated how motion and energy are related. Similar projects in which students are engaged in learning by applying science and engineering practices are generally appealing to funders.

Reason 3: To Launch New STEM Programs

Currently, there is great interest in improving science, technology, engineering, and mathematics (STEM) education and in increasing the number of students who are prepared to continue their education in these fields. One has to look no further than the White House's Educate to Innovate initiative (*www.whitehouse.gov/issues/education/k-12/educate-innovate*) to see that this focus is being taken seriously. In response to the call to improve the teaching of STEM disciplines, many schools are interested in offering additional science and engineering courses, Maker Clubs, and other opportunities for students of all ages. Obtaining grant money could facilitate this process and enrich your students' experiences. If you decide to pursue the process of obtaining grant money to start a STEM program at your school, you will need support from teachers and administrators, but the rewards and returns of such an endeavor will be tremendous.

Reason 4: To Collaborate With an Informal Science Education Organization

Science museums, state parks, national parks, and other nonprofit organizations are hubs of informal science learning that typically offer a multitude of programs that merge science content with real-world applications. Attendance at these programs, however, is not always free. Additionally, you may need funds for transportation to the site or for the purchase of consumable materials. Securing funds to work with these entities can be mutually beneficial and rewarding; very often students who visit science museums and parks on field trips develop long-term interests in the functions of these organizations and choose to volunteer or to become counselors for summer camps organized by these organizations. Needless to say, such activities help to channel student interest in science and promote their engagement in the work of community organizations.

Reason 5: To Fund Professional Development Opportunities

There are a myriad of professional development opportunities for K–12 science teachers. Although most occur within the United States, overseas programs also exist. Typically, the applications for such opportunities resemble a small grant proposal, requiring you to explain why you should be selected, the projects you will complete with your students after the training, and how your students will benefit from you attending the professional development activity. The key to being selected is your ability to outline and submit an application that is well supported with evidence that you possess compelling ideas regarding how to use this professional development opportunity to enhance your students' learning.

Reason 6: To Grow as a Professional Educator

Grant writing involves in-depth reflection and examination of your teaching and how it contributes to your students' learning of science concepts. As you justify

your ideas to a grant funder, you will find that the process of reflecting on your teaching will help you improve your practice. Although it is unlikely that every grant proposal you write will be funded, the growth you will experience as a result of the process will ultimately benefit your students.

Reason 7: To Participate in the Selection Process of Funding Organizations

Many organizations that fund educational projects require a teachers' perspective when deciding what proposals to fund. Sometimes simply contacting the organizations and explaining that you would like to volunteer as a judge will be sufficient to put you on the list of potential judges. It is not unusual for educators who have received grants or professional development from a funding agency to be invited to serve as judges in the selection of the next round of funded proposals. This process, which often involves readings and discussions with other passionate educators, will provide insight into a variety of classrooms and teaching practices.

Reason 8: To Receive Professional Recognition

Being awarded grant funds to implement innovative ideas in your classroom may propel you into becoming someone whom others seek out for expertise and advice. Your students and their parents, your colleagues, your school district, and possibly even your state may recognize you as someone who is passionate about enhancing student learning and who knows how to fund novel ideas. This type of recognition can be very helpful if you apply for a teaching award at the district, state, or nationwide level.

Reason 9: To Earn Professional Endorsement

Whether the granting organization is an educational one or not, obtaining external funds to implement your own classroom projects is a validation that your teaching practices, ideas, and goals have merit and are worth funding.

Reason 10: To Attain Personal Satisfaction

There is nothing like receiving a notification that your idea will be funded. It is both exciting and personally satisfying to know that your vision was successfully conveyed to the funder. Even more important, the opportunities that grant funds will bring to your classroom will reinvigorate your teaching, positively influence student learning, and increase your personal satisfaction with your role as a science educator.

Exercise

Think about your own motivations for writing a grant proposal. The list provided in this chapter is by no means complete because it is not personalized. Can you write one or more reasons why you will start writing grant proposals?

CHAPTER
2

Identifying and Refining Ideas for Potential Grant Proposals

You may be asking yourself, *What exactly is a grant?* A grant is the equivalent of a monetary gift that is given to the recipient in exchange for completing a specific project or other work that closely aligns to the mission and the goals of the funding agency. You can think of a grant as a contract between you and the funder. Although you don't have to pay the money back, you essentially are agreeing to carry out your proposal and ensure that your project adheres to the funding requirements specified by funding agency.

Grants are awarded by both nonprofit and for-profit organizations, as well as by local, state, and federal government entities that often have money designated to support specific types of educational opportunities. Most granting entities require that individuals seeking grants submit a grant proposal that outlines the idea and clarifies any points that the funding agency requires. Some grant applications are fairly complex, requiring a great deal of effort on your part, whereas other grant applications are somewhat simple, requiring only a few well-thought-out paragraphs. Regardless of the grant you decide to pursue, the key to successfully securing funding for your dream is for your passion to be clearly conveyed via your writing. This will require convincing the funder that your dream is one that will have a lasting effect on your life and on the lives of your students.

In our work with teachers during presentations on grant proposal writing, we have noticed that many wish to pursue a grant for the purpose of obtaining equipment or supplies. Although this is a commendable desire, we believe that your grant proposal will have an increased likelihood of being funded if you can strongly tie the need for classroom materials to student learning. Once you have identified a particular piece of equipment as being essential to your classroom, explain its relevance to your project goals. If you need to refine an idea that you currently have, or you have yet to identify an idea, you may find one or more of the following sources helpful in clarifying your thoughts so that your idea can be developed into a grant proposal.

Sources for Ideas
Your Students' Questions
Your greatest source of ideas is your students. Students are often capable of asking amusing and ingenious questions even though they may not always possess enough knowledge to know what will or will not work. Whether or not you feel comfortable allowing students to perform open inquiries, by eliciting student questions, you are validating their thinking and engaging them in the science and engineering practices outlined in the *Next Generation Science Standards* (*NGSS*; NGSS Lead States). Careful facilitation of classroom discussions can guide your students toward conducting inquiries designed to foster conceptual understanding or solve a community problem.

In addition to garnering ideas from them, you may want to consider enlisting your students to write portions of the grant. High school students can use this authentic experience to develop a student research project, while younger students can develop ideas for improving some aspect of their school community. If you like the idea of engaging students in the grant writing process, you may want to peruse the grant writing lessons found at *The Teaching Channel* (*www.teachingchannel.org*) and at *Learning to Give* (*www.learningtogive.org*).

Standards

You may want to begin the grant writing process by examining your district and state standards, as they may be helpful in guiding you in the selection of project ideas. Additionally, the *NGSS* performance expectations, with their clearly delineated connections to science and engineering practices, represent a wealth of potential project ideas. A quick perusal of the *NGSS* performance expectations for your grade level will no doubt inspire you to integrate experiences for your students that engage them in connecting their learning to the world around them.

Funded Grants

Although replicating a previously funded grant is not recommended, you may find that reading over successful grant applications will inspire you to integrate an original twist into an idea that has proven merit. For instance, the Toshiba America Foundation posts examples of previously funded grants for K–5 and 6–12 classrooms on its website. Reading the funded grant proposals may help you to clarify the type of grants that they are interested in.

Community Resources

The School Community

You will find that students are often more motivated to solve an actual problem that resonates with them than to conduct an investigation that is not well linked to a real-world setting. One way to pinpoint interesting questions for investigation is to facilitate a class discussion that deals with issues and problems that are typical for the school community. Prior to the discussion, the students can engage in conversations with the school nurse, principal, teachers of special needs students, cafeteria manager, and others by asking them to describe problems they encounter during the course of a typical day.

Your Local Community

Whether you choose to define your community as your school or choose to broaden your scope to encompass your school district's geographical area, there exists a wealth of resources you can use to foster and sustain your dream. Reading the local newspaper will often provide ideas for potential investigations that are closely related to the community. A plethora of topics, such as groundwater

contamination, land use, and pollution can serve as a catalyst for identifying investigable ideas.

If you are teaching environmental science or life science, you may find local, state, and national parks to be invaluable resources in generating ideas for student investigations regarding topics related to ecosystems and natural resources. You can open the door to an opportunity by making phone calls for networking with the park staff, whose goal it is to provide exciting educational experiences for students. Knowing the people who work in the educational outreach offices can place you on the top of the list when funds for transporting students become available. Even if you are in a small town or rural area, the internet makes it relatively easy to connect with individuals who may be willing to assist you.

Local Colleges and Universities

When exploring topics for a grant proposal, consider resources outside your school community. If you are fortunate enough to have a college or university in your community, reach out and make a connection. Universities and colleges are concerned with the low interest in STEM (science, technology, engineering, and mathematics) disciplines exhibited by undergraduate and graduate students and often have outreach programs in which you can request a contact with a scientist. Don't be intimidated at the thought of contacting a researcher, as oftentimes they must conduct outreach as part of their professional obligation to the university or to meet the stipulations of a grant that they have received. If you invite a scientist into your classroom, students will often ask numerous interesting questions that can turn into future grant proposals.

Establishing connections with research scientists in universities can also lead to exciting professional development opportunities for you, since some scientists will hire teachers to work in their laboratories during the summer. Working alongside a scientist is a great way to develop grant proposal ideas for your classroom projects. Ask to be added to the e-mail distribution lists of any local colleges and universities that sponsor ScienceCafés (*www.sciencecafes.org*), where discussions about science topics designed for the general public take place. Participation in these types of events will provide ideas for science proposals and allow you to interact with science professionals who may be able to assist you in your future work.

Your Business Community

Businesses are another terrific source of expertise and support; many find it difficult to turn down a teacher's request to help make learning science innovative and interesting. For example, as part of a grant-funded project, two tissue culture businesses in Florida supplied plants for use in a middle school classroom. In fact, the research and development director from one of the companies visited the classroom and taught the students how to pot and care for orchids. You can identify local businesses that may be willing to work with you by investigating your town's Chamber of Commerce.

Colleagues

Collaborating with other science teachers is a great way to extend learning opportunities to students outside your classroom, while collaborating with teachers from other disciplines can help you to develop interdisciplinary projects that tackle research problems from multiple perspectives. You don't need to limit yourself to colleagues within your own school or school district, however. Consider collaborating with teachers that you have met at conferences or through community groups via websites such as Edmodo (*www.edmodo.com*), Schoology (*www.schoology.com*), and Twitter (*www.twitter.com*).

Professional Organizations

Professional organizations, such as the American Association of University Women, the Institute of Electrical and Electronics Engineers, and the Society of Women Engineers, may also be able to assist with either generating project ideas or providing subject-area specialists who can lend their expertise to a project. Similarly, discipline-based organizations, such as the American Chemical Society, often have paid staff who can field questions, offer advice, and suggest project ideas.

Vendors

Review the catalogs of companies that supply materials and equipment for K–12 STEM education, such as Carolina Biological, PASCO, Fisher Scientific, Vernier, Texas Instruments, Ward's Science, and any others with which you may be familiar. You may find that reviewing the equipment and kits available for purchase can help you come up with an original idea for a project.

Your Personal Experiences
Personal Hobbies

Your own curiosity and interests can be another source for ideas. Do you have a hobby or passion that you wish to incorporate into your teaching? If so, consider bringing that idea into your classroom as a way to inspire your students. For example, birding is a popular hobby and one that can provide the inspiration for a project that involves students in observing birds and collecting data for scientific research, thereby building a connection between the real world and classroom activities.

Professional Development Experiences

Don't underestimate your own professional development experiences as a source of potential ideas. For example, attendance at the Science and Our Food Supply workshop, cosponsored by the National Science Teachers Association (NSTA) and the Food and Drug Administration, provided the spark for a proposal that addressed food safety and engaged students in learning how microbes are used to make a variety of foods. Similarly, a workshop held at a state park that featured descriptive signs along a hiking trail provided the impetus for seeking and obtaining funds to implement an interpretive walking trail located on school campus.

When attending professional development activities, engaging in conversations with other teachers and the workshop presenters can help you develop both ideas and partnerships. Whether your idea for a classroom project springs from a television show, a professional development session you attended, or a personal hobby, you have an excellent chance at convincing a funder to support your passion if you are able to link it to experiences grounded in real-world applications.

Exercises

1. What is your dream for your classroom, school, science department, or community? Brainstorm and write down ideas you would like to implement in your teaching.

2. Make a list of possible community resources that you could tap into for ideas and support.

3. Listed below are some areas that could be used to focus an investigation. Place a check mark next to those that appeal to you. Next to each idea, list one or more community resources that could support your idea.

Solar energy	Engineering design	Science and technology
Environment	Technology	Food science
Alternative energy	Physical science	Maker spaces
Recycling	Health sciences	Environmental engineering
Landfills	Earth science	Sustainability
Invasive species	Forensic science	Biotechnology
Nutrition	Conservation	Groundwater contamination
Erosion	Soil science	Connecting science and math
Robotics	Agriculture	Natural resource management
Programming	Gardening	Fuel efficiency
Endangered species	Air quality	

CHAPTER
3

Getting Started: May the FORCE Be With You!

deally, by now you've identified an idea that may rely on funding provided via a classroom grant. If not, we encourage you to do so because having an idea for a project will make reading this book and going through the exercises more meaningful and will result in a rough draft of a completed grant proposal.

The first thing you need to do prior to embarking on writing a grant proposal is arrange a meeting with your administrator(s) to explain your ideas and garner their support. Once you receive your administrator's approval to pursue a grant, you will need to begin the grant proposal writing process by considering how your idea is linked to your district and state standards and how it matches the funding agency's mission. Most school districts will require that student projects be aligned to school district, state, or national standards. Some granting organizations may require you to reference a specific set of standards, while others may not. Depending on the organization, the exact set of standards you use is probably not as important as linking student learning to your methods.

Figure 3.1 represents the intersection of the three major elements that should be considered: (1) alignment to standards, (2) the funding agency mission, and (3) your personal interests and interests of your students.

Figure 3.1

Successful Grants Represent the Intersection of Several Domains.

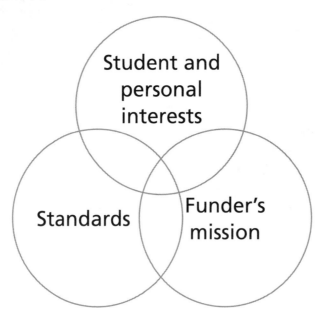

As you begin to settle on a specific idea, think about how it can connect to the *Next Generation Science Standards* (*NGSS*) and the *Common Core State Standards* and to science, technology, engineering, and mathematics (STEM) careers. Although grant reviewers may not be as familiar with educational terms as you are, aiming to meet new national initiatives will mean that your dream will be innovative and will use the latest research-based techniques to meet your students' needs. Grant reviewers will find it helpful if you explain any terminology in ways that a noneducator would understand, since those who read over your proposal may be unfamiliar with terms and acronyms such as *differentiation, benchmark, heterogeneous grouping, individualized education program (IEPs), English language learners (ELLs)*, and so on.

Grant Writing Is a Process

You may find the approach depicted in Figure 3.2 useful as you begin your journey toward a successful grant proposal. May the FORCE be with you!

Figure 3.2

Use the FORCE Acronym When Starting Your Grant Proposal.

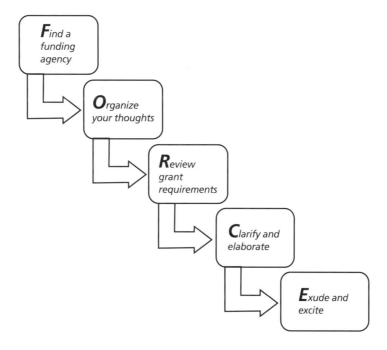

Find a Funding Agency

You can begin the process of locating a funding agency by carrying out a search on the internet. If you are a member of the National Science Teachers Association (NSTA), consider perusing the NSTA e-mail lists and reading the monthly *NSTA Reports* to identify sources of potential funding. Examining the websites of professional organizations, attending professional development workshops, and engaging in discussions with colleagues who have been successful in receiving classroom grants can also help you to locate funding agencies. Additionally, most states have grant programs; if you are interested in pursuing these opportunities, explore your state's department of education website. More information can be found in Appendix 6 (p. 133), which lists the offices of each state's department of education and U.S. territories. Once you've done your research, select the program whose mission best aligns with your project's objectives.

A Word About Obtaining Funding From Local Entities

When it comes to funding, never underestimate the power of your own community. Many school districts and/or parent-teacher organizations have small classroom grants available that have been established to help teachers obtain equipment or supplies that your budget may otherwise not cover. Parents themselves may have access to grants through their places of employment, and they are usually more than willing to help you obtain resources for your classroom. Businesses within your school district will often support you as well; stores such as Walmart, Lowe's, and Best Buy offer community and corporate grants to classrooms.

You may be surprised at how easy it may be to obtain equipment or materials without having to go through the grant proposal writing process. If you are fortunate enough to have a STEM business or college in your area, you may want to call and introduce yourself as a science teacher and inquire if any lab equipment is available for free. (Prior to accepting any equipment as a gift, make sure that it is safe and age appropriate for your students, and check that you have the permission of the school district to accept the equipment.) Having a specific project in mind that you can use to justify the need will help you promote your idea to the community. We have known teachers who have

- obtained balances from the local police station,
- been given free laptops from a computer store or company, and
- obtained digital microscopes from a hospital that was upgrading their equipment.

Organize Your Thoughts

There's perhaps nothing more intimidating than looking at a blank computer screen and wondering how to proceed. The methodology you employ to approach your writing is not important; what is important is that you get started and don't let the thought of the end product overwhelm you. You may find it helpful during this process to talk with other educators, parents of students, and your students. Sharing your ideas with others will help you identify what you want to convey to the granting agency. By including varying perspectives, you can broaden your scope and incorporate items you had not previously considered.

A proven method to start the grant proposal writing process is to simply brainstorm. Jot down everything you hope to accomplish should you be awarded the grant. Alternately, you can also use one of the following approaches when beginning the grant writing proposal process.

Create an Outline

You can create an outline using the grant components listed in Chapter 5 (p. 57) to construct your ideas, or you can use online software to create thought bubbles that can be arranged on your virtual canvas (search for "diagram software" to locate free online software).

Make an Elevator Speech

You may find it helpful to practice delivering an elevator speech by explaining your vision in the time it would take an elevator to climb several floors. Record your speech using your smartphone or other device, and then type it up or use speech-to-text software to transfer your words to paper. Don't worry about getting it perfect the first time, since the point is to get something on paper to start the process.

Begin With a Specific Proposal Component

Another approach is to start organizing your thoughts by beginning with the portion of the grant proposal that you feel is the easiest to complete. You may want to start by listing what you want your students to learn from the activities or investigations that you have planned for them. *Hint:* Since the focus of the majority of classroom grants is to improve student learning, listing *NGSS* performance expectations will provide you with a powerful place from which you can approach the other grant components.

Review Grant Requirements

After you have identified a funding agency for your grant proposal, carefully review its funding guidelines. Regardless of where the source of inspiration comes from, it will be critical to develop a proposal that is both matched to the funder's mission and aligned with your school's demographics. You should be aware that some grants may be targeted specifically for urban schools, underperforming schools, or Title I schools, while other grants may be designated for a specific scientific discipline. For example, the Captain Planet Foundation funds environmental projects, which can be broadly defined. Although many topics could be related to studying the environment, it might be difficult to obtain funding for a project related to exploring the relationship between building materials and the design of cars, which falls within the scope of engineering sciences and technology. Even though cars can affect the environment, if the project does not address this aspect, the project will not be funded. It is therefore critical to review what the funding agency expects from you and to match your needs to their mission.

Most agencies have specific guidelines that accompany their application forms; it is crucial to pay close attention to the stipulated guidelines to ensure that your proposal will be considered. You may find a great deal of variability between funding organizations, with some asking for responses to a few questions and others asking for a detailed proposal that contains all the components described in the following chapter.

> **Tip!** You may want to consider sending your grant proposal to more than one funder, provided you carefully follow the requirements specified by each of the funding organizations.

Clarify and Elaborate

Although grant applications can vary tremendously between organizations, all funders have one basic requirement: that your idea be clearly and concisely presented in a way that communicates its value and your ability to complete the proposed work. This may require several drafts on your part, as well as the assistance of an editor. Writing concisely can be challenging, but is crucial for clearly and effectively conveying information. There are numerous web pages devoted to tips that will help you write more concisely. Alternately, you could ask an English teacher or someone else with writing expertise in your school for assistance.

Pursuing a monetary gift will involve time and effort on your part to complete the grant application. When writing the grant proposal, it is crucial to convey your passion for the proposal idea and effectively explain how it relates to your students' learning and appreciation of science. One method for approaching your first draft is to allow the fervency for your ideas to come forth in a stream-of-consciousness approach, writing anything that comes into your mind regarding the grant proposal. When composing the first draft, ignore issues related to spelling, grammar, and word flow. Once you have your thoughts down, you can correct the spelling, clean up the grammar, and rearrange sentences and paragraphs so that they make sense. Stopping along the way to make those corrections slows the process and can be frustrating and inefficient.

Probably the biggest obstacle you will face in writing the grant proposal is the tendency to procrastinate. Key to staying motivated throughout the process is to remember why you are writing in the first place. As you think about writing the grant proposal, tailor it for your specific setting and learners. For example, you may be interested in providing enrichment experiences for your gifted students, supporting ELLs in your classroom, or promoting learning in students who possess learning disabilities. It is a good idea to start the writing long before the grant proposal submission deadline and to work on it consistently rather than in binges. Although binge writing can result in successful grants and awards, many authors will tell you that is not as effective as setting aside time daily to work. It is always a good idea to plan on submitting your grant proposal a week prior to the due date. That way, if you run behind schedule or have difficulty obtaining documentation, you will have plenty of time to complete the proposal.

Once the proposal is finished, ask a friend or someone who does not teach science to proofread the work prior to submitting it to the funding agency; he or she can evaluate it using the "Grant Proposal Rubric" in Appendix 2 (p. 117). If this person can understand your vision, then you did a great job communicating it!

Exude and Excite

A prerequisite for receiving funding for your idea is the ability to demonstrate confidence in presenting and carrying out your proposed concept, as granting agencies want to know that their funds are being awarded to an individual who can complete the work described. Provided that your confidence comes across in your writing and that you have developed the grant proposal to align with the funder's mission and proposal guidelines, your ideas have an excellent chance of

> **Tip!** Remember to include the school administration early on in the writing process. Administrative support is vital to the success of any bold endeavor, given that your "dream" may necessitate altered schedules, field trips, guest speakers, and more.

being funded. The next chapter will take you step by step through the process of developing the most common proposal components that funding agencies require of K–12 classroom grants.

CHAPTER 3

Exercises

1. Select at least two funding agencies that you can submit your grant proposal idea to.

2. Identify the missions of the agencies and outline their major requirements for the grant proposal application.

3. Think about how your idea will fit with the missions of these funding agencies.

4. How you can connect your dream to your state standards and/or to the *NGSS*?

Grant Proposal Components

CHAPTER 4

This chapter, which addresses the nuts and bolts of grant proposal writing, is broken down into sections that target the typical components required by many funding organizations. Although each proposal will vary depending on the granting organization, many organizations require the components depicted in Figure 4.1 (p. 24).

Each component shown in Figure 4.1 is discussed in-depth in relation to K–12 science education. Tips for writing each component, examples from funded science grants, and helpful exercises have been included to guide you through the grant proposal writing process. Corresponding templates designed to help you organize your thoughts and begin your writing can be found in Appendix 1 (p. 101).

As you progress through this chapter, you may be surprised (and somewhat relieved!) to see that many typical grant proposal components strongly resemble

Figure 4.1

Standard Grant Proposal Components

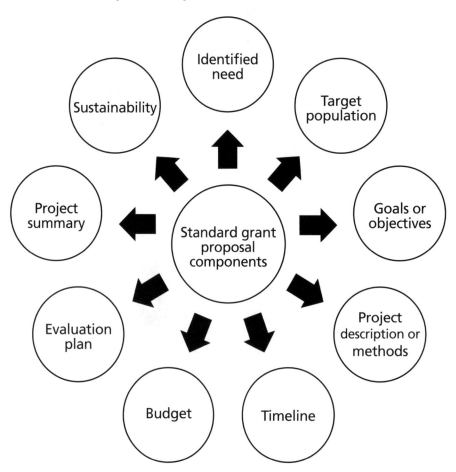

aspects of a lesson plan or are activities that you already have experience conducting. It helps to take the intimidation factor out of grant writing when you realize that *you are already an expert at several grant proposal components.* The important thing is to not get overwhelmed by the process, but instead to simply begin getting your thoughts down on paper. Although it doesn't technically matter which component you start with, a logical place to begin is by identifying your objectives because they will be critical for establishing your grant proposal's framework. Funders want to know that they are positively influencing students. Even if your proposal is written mainly for the purpose of securing professional development or materials, it is critical to address how your request for funds is linked to enhancing student learning. Finally, and perhaps above all else, let your passion shine throughout your writing. Your proposal should convey your enthusiasm for your idea and your confidence in carrying it out. What is *your* dream? What are *you* passionate about? How can you bring that into your classroom? If you can convey excitement about your proposal and the power to affect your students is clear, people will *want* to provide you funds to carry out your dream.

Are you ready? Let's get started!

Describing the Need and Potential Impact

The needs statement, or rationale, is the place where you can communicate your passion for your dream. Defining the need will also help you to clarify in your own mind exactly why you want this project to be funded. The statement should be written in such a way that it persuades the funder that your proposal can effectively address the need. Your identified need may be restricted to your students, or it may be broader, incorporating the school district or perhaps your entire town, as show in Figure 4.2. In most cases, the needs statement will not be more than one or two pages long. To prepare a well-organized needs statement, begin by answering the questions listed on pages 26–28.

Figure 4.2

Grants Address a Variety of Needs.

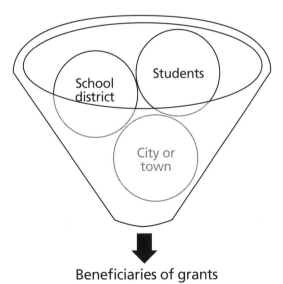

Beneficiaries of grants

What Problem or Need Exists in Your School, School District, or Community?

Start by identifying the problem or need that exists in your community. How did you realize that a need existed within your school, school district, or community? Was it through an observation or conversation with a student, colleague, or parent? Perhaps it was the result of a hobby—something that you would like to parlay into a learning experience for your students. For example, if you are a gardener, is there a way you can ignite that same passion for gardening in your students while providing them with hands-on opportunities to make connections about soil, nutrients, plant requirements, and agriculture? Do your students struggle with physics concepts? Could you use your passion for skiing to create engaging lessons about force and motion?

Perhaps the identified need has arisen from the community in which your school is located. Is your local community struggling with an issue related to natural disasters, such as flooding and tornadoes? If you are in a farming community, are drought, erosion, pesticide use, or invasive species a concern? If your school is located in a city, perhaps traffic patterns, air quality, or waste management are topics worthy of investigating. Linking your proposal to real-world problems and events helps students make better connections to concepts and strengthens your proposal (more information about how to get started identifying an idea can be found in Chapter 2, p. 7).

What Opportunities Will Your Proposal Provide for Your Students, and How Will Your Students Grow?

It is critical that your proposal explain how the project will positively influence student learning, even if the proposal is written for the specific purpose of obtaining laboratory equipment or supplies. Consider focusing your needs statement on how the equipment will enhance the learning that will occur in your classroom and less on the equipment itself. You may find it helpful to think about how the equipment you are requesting can be used in a manner that will engage your students in the science and engineering practices identified by the *Next Generation Science Standards* (*NGSS*) rather than simply to support verification labs.

What Is Your Evidence of Need?

The use of summary statistics related to Title I numbers or English language learner (ELL) populations is appropriate when describing your school or school district's need, as long as you keep your descriptions brief (given that there are many schools with large numbers of underserved students, try to keep your focus on what makes your project worthwhile, rather than relying purely on need). Provided you keep within the word limits set by the granting organization, you may want to consider supporting the need for the proposed project with information from Appendix C of the *NGSS*, references from a research study, or information from one or more of the National Science Teachers Association (NSTA) position statements.

> **Tip!** Avoid using circular reasoning, in which the need and the solution are the same. For example, perhaps you would like to develop an outdoor classroom for your school. Although this is a very worthy cause, be careful when outlining your justifications. Rather than stating that you need an outdoor classroom and that building one will address the need, *frame the need in the terms of student learning*. In this scenario, explain how the establishment of the facility will engage students in learning opportunities grounded in an authentic context and deepen their connection to the natural world.

How Will Your Proposal Address This Need?

Next, discuss how your proposal will address the identified need. Describe the major goal or goals of your proposed project, and explain how your students and (if applicable) the local community will benefit. It should be evident when reading the goal(s) that they are clearly linked to the need.

Who Will Be Affected by Your Project?

State the approximate number of teachers and students that will be affected by your project. You may also want to include a brief description of your students that includes grade and ability levels. Don't worry too much if your project involves a small number of students because it is often the strength of your proposal rather than the sheer number of students that will sell your idea to the funder. If your project will be conducted at the district level, it is a good idea to also include a description of the students in the district who will benefit from the project.

Needs Statement Example

Below is an example of a needs statement from a funded grant. Notice that although the proposal was written for a professional development opportunity for the teacher, the need is connected to student learning. Depending on your proposal and the specific grant you are targeting, your needs statement may be considerably longer (up to two pages).

> *Perhaps no other topic in ecology is as fascinating to my students as the topic of rain forests, yet I have no field experience or formal training in tropical ecology. I believe that students learn best through field-based, inquiry-driven curriculum. This philosophy is supported by the NSTA "Position Statement on Scientific Inquiry," which recommends that science teachers "plan an inquiry-based science program for their students by developing both short- and long-term goals that incorporate appropriate content knowledge."*

> *Attendance at the Summer Institute Connections: Tropical and Temperate America will provide me with opportunities to network with teachers, scientists, local biologists, and archaeologists involved in conservation projects. It will also provide the background necessary to develop field protocols that will engage my students in using tropical forest field techniques to monitor deciduous forest plots on our school property.*

Important Points to Remember

- Frame the need in terms of student learning.
- Let your passion for your idea shine through your writing.
- As with the other portions of the grant proposal, write concisely when composing the needs statement.
- Use data related to your students or school district to sell your idea to the funder.

Exercises

1. What needs exist in your community that could lend themselves to a potential grant proposal?

2. What statistics or evidence do you have that can best support your identified need?

Your Turn!

Using the "Writing the Needs Statement" template in Appendix 1 (p. 103), describe the need and potential impact of your project.

Describing the Target Population

Granting agencies want to know who is benefiting from their support, with most requiring accurate demographic information. Use the "Documenting Target Population Demographics" template in Appendix 1 (pp. 104–105) to guide the data-mining process. You should consider including the levels of information listed next in your description.

Demographics of Your Students or Classroom

There are a number of factors that can be used to describe the students in your classroom. At a minimum, you should include the total number of students benefiting from your project and their age or grade level, and the percentage of minority students, ELL students, gifted students, and special education students. If appropriate for your setting, you could describe the number of students who may be economically disadvantaged or the gender ratio of your class, particularly if your proposed project is designed to encourage minority and female students to pursue STEM (science, technology, engineering, and mathematics) careers.

Demographic Makeup of Your School District

Incorporating information about your school district helps grant readers visualize your setting. When deciding what information to include, *make sure the data help illustrate the need*. Think about any compelling statistics that will convince a funder that your proposal will benefit a needy group of students. If your district has a large underserved population, you can help justify the need through the use of statistics such as the percentage of students who come from single-parent homes, are receiving free and reduced lunch, or are attending Title I schools. Other school data that you may be able to use to illustrate the need for your project include state test scores, graduation rate, dropout rate, or the percentage of students who enroll in AP courses. If you have your eye on technology equipment, consider including the student-to-computer ratio.

Demographic Makeup of Your Town or City

Demographic information about your school district's residents can also be used to convince the funder of the existence of

> **Tip!** Demographic data change; make sure that yours are up to date when you apply for grants.

needs in your community that have not been met. This information could include the population and ethnicities of people who reside in your school district, the median income, and the percentage of people who live below poverty level. If appropriate, consider comparing the median income of your school district to the median income of your state. The information will be particularly compelling if your district or locale includes economically disadvantaged or minority groups. That is not to say, however, that more wealthy schools or schools with homogeneous populations won't qualify for grants.

Important Points to Remember

- Locate data to use in your description.
- Decide which data will best convince the funder of your need.

Exercises

1. Can you think of any data you could incorporate into your proposal to support your claim that a specific need for the project exists in your classroom, school district, or community?

2. Locate your school at the National Center for Education Statistics (*www. nces.ed.gov*). Explore the site and select data you could use to support your grant application.

3. Examine the demographic data described below, which was used for a funded grant. If this were your school community, what additional data could you incorporate to support evidence of need?

   The target student population will consist of two seventh-grade teams. A total of approximately 300 students will be participating in the activities. The student population will consist of regular education students and students who have been identified as needing learning support, emotional support, or gifted support. Approximately 90% of the participants are Caucasian. The remainder of the participants include students whose background is Asian or African American heritage.

Your Turn!

Use the "Documenting Target Population Demographics" template in Appendix 1 (pp. 104–105) to gather demographic statistics related to your classroom, school district, and community.

Developing Your Project Goals and Objectives

The process of writing your grant proposal objectives is very similar to what you may use when writing a learning goal or lesson objectives. There are a few things to keep in mind, though. First, think about why you are writing the grant proposal. What is it that you hope to accomplish, and how does it relate to student learning? How is obtaining the grant going to help you to successfully meet your project's goals? Second, the adage "less is more" is an apt one when it comes to writing your project's goals. Although it may be tempting to write numerous goals, you will have a greater likelihood of success if you concentrate on writing a few well-thought-out objectives rather than a multitude of items that are not closely linked to the identified need. It is also important to remember that objectives are not activities. Rather, most grant proposals require a separate methods section where you explain how you intend to accomplish the identified objectives.

SUPER Goals

Grant proposal readers appreciate objectives that are concisely written and leave no doubt about what you are hoping to accomplish. Once you have identified the

> **Tip!** Make sure your project objectives align with the funder's mission and/or standards.

learning goals, you will need to write the objectives in a manner that makes them easy to understand and visualize. When writing these objectives, make sure they are SUPER (see Figure 4.3, p. 32).

Specific. Specificity leaves no doubt in the funder's mind about precisely what you hope to achieve with your grant. What do you hope to accomplish? Who is your target audience?

Use science and engineering practices. When writing your goals, employ the most current science teaching practices to demonstrate that you are up to date regarding science education.

Practical. Your goal should be achievable and age appropriate. You are an expert in both your content area and pedagogy. As such, you know the type of developmentally appropriate activities that will stretch your students.

Evaluable. How will you know that you have met your objective? What will be the outcome after the grant proposal has been implemented? What will students be able to accomplish as a result of your proposal?

Relevant. How is your goal significant to your defined community, and how does it relate to student learning?

Figure 4.3

SUPER Goals

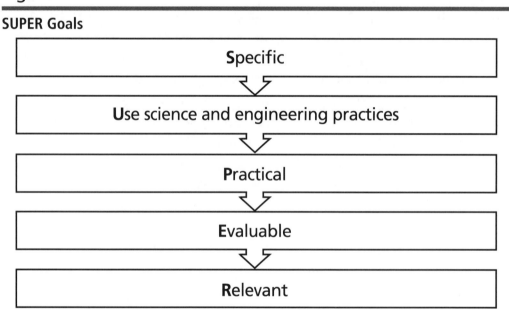

Important Points to Remember

- Write concisely.
- Limit the number of goals.
- Write SUPER goals.

Example Goals and Objectives

The following objectives were developed for a grant that involved middle school students in identifying breeding birds for the purpose of contributing to a state *Bird Breeding Atlas*. Do you feel that they are SUPER objectives?

> *As a result of this proposal, all students will be able to carry out the following actions by the end of the birding unit:*

- *Focus binoculars properly using diopter adjustment*
- *Identify 25 common birds by sight and sound*
- *Develop science process skills, including classification, observation, data collecting, and analysis by keeping a birding journal in which date, time, weather conditions, location, effort (time spent watching), and detailed observations are recorded*
- *Correctly characterize habitats based on vegetation and physical geography and predict common species that may or may not be present in a given habitat*
- *Use transects and point counts to census birds within different habitats*
- *Use Excel to contribute to a database that includes latitude, longitude, and species of bird observed*
- *Construct graphs of data and compare to historical records*
- *Use central means of tendency statistics (mean, median, and mode) to analyze data and to draw conclusions*

Exercises

1. What are the components of a SUPER goal?

2. Rewrite the following goal into a SUPER goal: Students will learn how to analyze water quality with test kits.

Your Turn!

Use the "Writing Your Goals and Objectives" template in Appendix 1 (p. 106) to write down your project goals and objectives.

Project Description (Methods)

Teachers of science are well versed in writing the project description section, which is also often called the methods section, as it is the equivalent to the instructional part of a lesson plan. This is where you tell the funder about the strategies and activities

you will use to address the identified area of need. This is your opportunity to show that you know what you are doing! You will need to write clearly and concisely as you did when writing the other grant components, taking care to avoid terms and educational jargon that a noneducator would not understand. As you write, try to connect the passion you have for your project to student learning.

Your methods section is a great place to show how you will engage your students in science and engineering practices such as planning and carrying out investigations, analyzing and interpreting data, and engaging in scientific argumentation using evidence. Deciding on which actions you and your students will undertake as part of the project can be challenging, however. One of the benefits of being a science teacher is that there are nearly limitless investigations, projects, and community service opportunities in which to engage your students. This can be overwhelming, which is why it is important to narrow your scope and focus on your objectives. To prepare this section in a way that best represents your project, start with rereading your proposal's objectives. Identify learning experiences that you feel will best meet the objectives, making sure that all of them are tied directly the identified need.

As you write your methods section, you may find yourself tempted to add in a lot of experiences or inquiries, but try to be concise and focus on the most important activities. You may also find yourself tempted to include experiences that are not age appropriate for the involved students, but it is recommended that you carefully align the activities to the abilities of the particular classroom and students. Aiming too high in your expectations may result in a mismatch between pedagogy and student learning. Finally, list all the events related to achieving your objectives or goals in sequential order; they should fit together in a way that clearly supports student learning.

Although your objectives will be educationally beneficial and designed to increase student learning, it is important to note that the educational benefit does not have to be restricted to having your students demonstrate cognitive growth. You can also include methods that will address the affective domain. For example, if your proposal involves improving STEM experiences for minorities and girls, it would be appropriate to include goals that encompass factors such as motivation. Likewise, an environmental science project that immerses students in their community may promote interest in STEM careers or heighten awareness of environmental issues.

You can provide a rationale for all of the learning experiences in the project description section by explaining how your proposed methods meet learning objectives or other goals better than traditional methods do. Similar to the other grant components, discuss how these experiences will affect student learning, deepen conceptual understanding, and help solve a community problem. Think about how guest speakers, field trips, and special schoolwide events that you may plan can enhance your students' learning experiences and community or parent-student-school relations.

Features of a Good Project Description Section

Exhibits Strong Connections to Student Learning

When applying for a grant from any organization, it will be critical for you to link student learning to the funder's mission. Regardless of which grant you pursue, all activities need to be age appropriate. If you are describing an activity in which you will be using equipment, focus on a particular learning experience and the inquiries that students will carry out rather than simply listing what students can do with the equipment. Some grants may require you to link student learning to a specific sets of standards and others may not. Keep this in mind if you are applying for a state or federal government grant, because in that case, you may be required to connect your methods to standards.

Incorporates Science and Engineering Practices

When planning the methods you will use to meet your objectives, it can be helpful to differentiate between doing an "activity" and "inquiry." Typically, with an activity the results are known ahead of time. Think of the verification labs that you may have had your students conduct in the past. Although many activities are excellent and are designed for the purpose of allowing students to master conceptual information through their use, the best way to engage your students in the *NGSS* science and engineering practices is by developing inquiries, that is, innovative, meaningful experiences grounded in real-world data collection or exploration. Many organizations that fund classroom grants do so for the purpose of getting students excited about science and thus tend to be impressed with those proposals that are based on solid science, where students are acting as scientists and addressing real-world problems using scientific way of investigation and thinking.

Includes Partnerships

Don't be intimidated by this suggestion: Although including a community partner can be beneficial to your grant proposal, it is by no means mandatory. That said, partnerships can add incredible depth and dimension to a proposal, so it is a good idea to make an effort to include experiences that involve collaboration with universities, colleges, state agencies, and local businesses. Partnerships can include mentoring opportunities within your school and your school district if other teachers and students from different grades or schools become involved in the project. Partnerships don't have to be restricted to your geographical area; you can try connecting with scientists and organizations through Twitter, Skype, or Scientists Twibes. You may also want to consider connecting with other classrooms via Edmodo, Schoology, or another online learning management system so that students can share any results with other students.

Involves an Authentic Audience

Consider having your students create a presentation, video, or paper that summarizes what they have learned through participation in the project if it is funded. You will find that the students will be motivated to do a great job by the possibility of presenting their findings to an authentic audience. Whether it is sharing water-quality data with township supervisors, mentoring younger students, or creating a public service announcement about an issue central to your proposal, the element of educating others is one that is valuable and inspiring.

Important Points to Remember

- Select age-appropriate activities.
- Limit the scope of your activities.
- Incorporate student-driven inquiries rather than "activities."
- Connect your passion for the project back to your work as a science educator.

Example Project Description

This project will involve students in the planning, digging, planting, tending, and harvesting of a community garden. The standards-driven activities will be hands-on and will help to develop an appreciation for the land, an understanding of the importance of recycling, and a sense of community pride within each student. Visitors to the school will have an opportunity to view the community garden, further widening the project's impact on the community. Specific activities that will supplement the gardening experience will include the following:

- Comparisons of two indoor composting systems—vermicomposting and aerobic composting. Comparisons will include data collection, graphing, and analysis of the two systems.

- Seed germination

- Planting and tending of vegetables. Students will understand the specific requirements of vegetables planted. Students will also learn how to identify weeds and how to assess the health of garden plants.

- Integrated pest-management activities. Students will learn how to identify harmful and beneficial pests. Students will study the life cycle of a lady beetle. Lady beetles will be raised and released into the garden to control insect pests.

- Harvesting garden crops. A record of all harvested vegetables will be kept. Vegetables will be donated to a local soup kitchen.

- Outdoor composting of garden organic material.

- Recycling of composted materials. Students will use composted materials in the garden to fertilize the garden organically.

Students will communicate their gardening success in a variety of ways, including posting articles and photographs about the garden on the school district website, writing articles for the school district newsletter, and acting as mentors to younger brothers and sisters who wish to start a gardening project at home.

Exercises

1. A typical methods section might be anywhere from one to three pages in length. Examine the sample. How does the author

    - focus on student learning?

    - engage students in inquiry science?

    - incorporate partnerships?

    - include an authentic audience?

2. A teacher submits a grant proposal to purchase a greenhouse, stating that her students will be using potting mix to grow vegetables. How can the teacher incorporate science and engineering practices into his or her proposal?

Your Turn!

Use the "Writing the Methods and Activities Section" template in Appendix 1 (p. 107) to write your project methods or description.

Timeline

Although treated separately here, the timeline is often an extension of the project description or methods section. Think about when you plan a unit: How do you decide the length of time it will take to teach the unit? If you apply the same approach when writing the grant proposal, you will no doubt experience success when creating the proposal's timeline.

When writing your timeline, begin by aligning the project's major events to specific months. This is not only easier and more practical than trying to tie your events to a specific date, but it also builds flexibility into your timeline. As a teacher, you have probably experienced situations in which things took longer to cover than expected. Unanticipated alterations to the schedule, such as snow days, hurricane days, assemblies, and parent-conference days can wreak havoc with your planning. As such, give yourself plenty of time to complete the experiences and investigations included in your project.

When writing the timeline, you may want to use the fiscal year as a guide, since most classroom grants are designed to start and end within the same school year (for most school districts, the fiscal year roughly corresponds with the academic year). One approach for writing the timeline is to work backward, listing all required tasks and the outputs that will be generated. Another option is to begin the process by brainstorming for all the activities that will need to be conducted to ensure that you maximize the success of the proposal. Be sure to include items such as working with the administration, making purchases, organizing field trips, and arranging for guest speakers and special events. If applicable, you'll want to also include any activities related to curriculum development and any evaluation and dissemination activities. If you are working with a partner organization or with a colleague, be sure to consult with them to ensure agreement on the timeline. You do not have to provide an extremely detailed list; one that includes the major events and activities is sufficient for most grant applications.

You may have the impression that the student learning experiences described in your grant proposal should cover the course of the entire school year. Although this is not an atypical approach, there is nothing about proposal writing that requires you to implement such a project. Your proposed idea can cover the course of the school year, or it can take place during a specific unit of study that is only six to eight weeks long. Remember that it is the overall strength of your proposal and the passion that you have for your idea that will fan the flames of interest among the grant reviewers rather than when and how much time the students will spend immersed in learning experiences.

Important Points to Remember
- Consider working backward or use brainstorming to plan for activities related to your proposal.
- Assign each major event to a specific month.
- Build in a buffer since things always take longer than you anticipate.

Example Timeline
Examine the example timeline or a grant that involved the construction of an interpretive trail on school campus, found in Table 4.1 (p. 40). Note how major events are summarized by the month. Unless required as part of the grant application, there is no need to go into a great deal of detail regarding actions since all learning experiences should be thoroughly addressed within the methods section of the grant.

Table 4.1

Example of a Timeline

TIME FRAME	ACTION
SUMMER	Teacher purchases equipment.
SEPTEMBER	Teacher trains students in field techniques such as how to use pitfall traps, measure soil chemistry, use dichotomous keys to identify trees, use increment borers, and take diameter at breast height and tree height.
SEPTEMBER	Students collect and press plants. Identify for herbarium.
OCTOBER	Students design an inquiry-based project.
OCTOBER	Students conduct projects and present results at a science seminar.
APRIL/MAY	Students research topics.
APRIL/MAY	Students use iPads to create podcasts.
APRIL/MAY	Students use iPads to take video and still pictures.
APRIL/MAY	Students use GIS (geographic information system) units to create a walking path and select areas of interest.
APRIL/MAY	Students create interactive Google maps.
APRIL/MAY	Students assemble a website.
APRIL/MAY	Students create brochures for an interpretive trail.
MAY	Several students present project at a home and school association meeting and at a school board meeting.
MAY	Students host a "trail day."

Exercises

1. Referring to your methods section, brainstorm to make a list of all events that will be carried out as part of your grant proposal. What other actions will you need to ensure that your grant is successfully completed?

2. Create a sample timeline for the example provided in the project description or methods section (e.g., a teacher submits a grant proposal to purchase a greenhouse, stating that her students will be using potting mix to grow vegetables).

Your Turn!

Use the "Creating the Timeline" template in Appendix 1 (p. 108) to create your timeline.

Developing the Budget

The budget, perhaps the most familiar of the grant proposal component pieces, can be written at any point in the grant writing process. You should be aware that many grant proposal reviewers peruse the budget first. What do grant reviewers typically look for when they review a budget? They examine it for any costly items and evaluate how well aligned the items are to the objectives of both the granting organization and the grant proposal. As such, it is vitally important that the budget be tied closely to the proposal's objectives and to the budget guidelines of the funding agency.

Some grants place limits or stipulations on what can be purchased. It is absolutely critical that you carefully read the grant guidelines to determine what the funders will support and what they won't. There is a great deal of variation between funding agencies; some organizations pay for stipends and curriculum writing, while others do not. Many organizations will not fund computer hardware purchases, but some do. Still other funders may have established limits on the percentage of the grant that can be used for specific items. Whatever the case, make sure that your budget meets the criteria set forth by the granting organization. As you can see from Table 4.2 (p. 42), which depicts information from several K–12 classroom grants, the Association of American Educators classroom grants can be used to purchase computers, but Toshiba American Foundation Grants cannot be used in this manner. Still other grants, such as the Project Learning Tree GreenWorks! grants have stipulations related to teacher attendance at workshops and require in-kind donations by the school district. The variability in stipulations set forth by the funding agency makes it crucial to match the proposed budget items to the funder's requirements and mission.

Strategies for Writing the Budget and the Budget Narrative and Justification

Many novice grant writers are under the impression that their proposal will receive higher consideration if they do not ask for the full amount of the grant. You should be aware that this is not the case. Granting organizations generally establish a specific sum that they wish to use for science education. If a block of $100,000 has been set aside for 20 grants of $5,000 each, the organization may end up funding well over 20 grants if the average grant request is only $4,000. You may want to create a budget for the full amount, *provided that what you are asking to fund is integral to*

TABLE 4.2
Budget Restrictions and Allowances From Sample Grants

GRANT PROGRAM	BUDGET TOTAL	ITEMS ALLOWED/NOT ALLOWED
ASSOCIATION OF AMERICAN EDUCATORS CLASSROOM GRANTS	Maximum $500	Funds allow for books, software, calculators, audiovisual equipment, lab supplies, and materials.
TARGET FIELD TRIP GRANTS	$700	Funds are to be used to cover field trip costs and fees. In the event that the costs and fees are less than the grant amount, the balance of the grant may be used for other education costs such as materials, books, and resources related to the curriculum. Funds may not be used for expenditures that are the normal responsibility of the school district (e.g., substitute teacher salaries).
TOSHIBA AMERICAN FOUNDATION GRANTS	K–5: Up to $1,000 6–12: Up to $5,000	Funds cannot be used for computer equipment, afterschool programs, textbooks, or audiovisual equipment. For a complete list of restrictions, see the grant website.
THE AMERICAN INSTITUTE OF AERONAUTICS AND ASTRONAUTICS FOUNDATION CLASSROOM GRANTS	Up to $250	Funds can be used for materials for projects related to aerospace science.
CAPTAIN PLANET FOUNDATION	$500–2,500	At least 50% matching in-kind funds have to be secured to be considered for funding. Funds cannot be used for operating expenses, salaries, scholarships, landscaping, building improvements, or promotional items.
VOYA UNSUNG HEROES AWARD	$2,000	Funds must be used to further the projects within the school or school system. Indirect costs or administrative fees should not be paid or withheld from the grant award.
PROJECT LEARNING TREE GREENWORKS! GRANTS	Up to $1,000	Applicants must have attended a Project Learning Tree workshop, and the proposed project must involve student learning, student voice, and a community partner. In addition, applicants need to secure at least 50% of in-kind contributions, such as volunteer time and donated materials.
LOWE'S TOOLBOX FOR EDUCATION GRANT	Between $2,000 and $5,000 per school is available	Funds are not to be used for a variety of activities, including student trips, private schools, and stipends. See the grant website for the full list.

your vision. If you truly only need $4,000 worth of equipment and decide to include a nonessential $1,000 whiteboard to round out your grant, you may run the risk of having your otherwise great idea rejected, as grant reviewers can effectively identify those proposed purchases that are not essential to the grant itself.

Funding agencies will sometimes require you to write what is termed a *budget narrative*, which is really nothing more than a justification of your proposed expenses. Although writing the budget narrative is fairly straightforward, you may want to write it after you have defined your proposal's objectives and identified those items that are integral to its success. Once you have done that, writing the budget narrative can be approached as follows:

- Research and obtain prices for the items required to carry out your proposal.
- Group items by category.
- Justify why items are needed for your proposal.
- If required, determine what in-kind services you will list.

When researching prices for the items you wish to purchase, it is a good idea to compare prices from a number of vendors, since prices can differ substantially. In your narrative, let the grant reviewers know that you shopped around for the best prices. Clearly identify the vendor, name of product, description of item, and the estimated cost. It is important not to underestimate the cost of your purchases because you may have difficulty carrying out your proposal if you don't budget carefully. Don't forget to calculate in shipping and handling, too, or you may end up with a heftier bill than anticipated.

In-Kind Services

In-kind services can include items such as donated time, materials, and services. The following list gives you an idea of common in-kind services:

- Your volunteer hours to write curriculum, order supplies, or anything else you may have to do to ensure the success of the grant
- Use of the school district classroom space over the summer months. (It is advisable to check with your school district and other entities prior to listing any in-kind service related to the use of their facilities).
- Donations from your parent-teacher association, school board, principal's fund, or other entity that is contributing money or supplies

Tip! You may be asked to supply proof of your school's tax-exempt status. Nonprofit organizations, or 501(c) organizations, are those that qualify for tax-exempt status. When an organization is granted 501(c) status, the Internal Revenue Service issues a determination letter that lists the tax-exempt status. Your school's business office should be able to easily provide you a copy of this letter.

- Donations of money, supplies, or time from parents or local businesses

- The use of facilities that waive a rental fee, such as your school gymnasium, township building, or public library

- Any donated advertising

Listing in-kind services donated by your school, parent group, or local business conveys to the grant readers that your proposal is one that the community supports. By the same token, donating your time to write curriculum or attend trainings indicates your commitment to the project and your willingness to ensure that it is carried out successfully. Although some grants allow you to write in a small stipend, donating your time will allow you to purchase items that can be used to successfully carry out your dream.

A Word About Spending the Grant Money

Some granting organizations will ask you to address project management in your proposal. As such, be prepared to explain how the funds will be secured and who will have access to them. Have a discussion with your administrator in which you clarify how money from a potential grant will be set aside. Most school districts will set up a special account as required by your state's auditing policies. Once you've gotten the grant money, you may have to go through the district to purchase your supplies. If purchasing supplies on your own, make sure that the vendor will take your school's 501(c) tax-exempt identification number to avoid paying sales tax. (If you purchase the items yourself and pay taxes on them, you run the risk of not getting reimbursed.)

Example Budget for a Large Grant (More Than $1,000)

Take a look at Table 4.3, which shows a budget for project that involved students in investigating the habitat preferences of woodland salamanders. You'll notice that even though the grant could have been for up to $10,000, the author opted to only ask for $9,500 and chose to include a stipend for the purpose of writing curriculum. Although the grant was funded, a wiser choice might have been

TABLE 4.3
Example of a Budget for a Large Grant

ITEM	ESTIMATED COST	JUSTIFICATION/BUDGET NARRATIVE
MAJOR EQUIPMENT SUPPLIES		Project expenses can be broken down into three major categories: technology equipment, lumber for cover boards, and technology training and stipend for project director. Handheld collection tools, sensors, and the digital camera will be used to collect data related to habitat preference of *Plethodon cinereus*. The laptop, which can be used in the field, is necessary to compile a database of information and provide students with a vehicle to generate Excel graphs. Lumber, spray paint, and flagging tape will be used to construct cover boards and to establish study sites.
Handheld data collection tool—10 at $150 each	$1,500	
Light meter—10 at $59 each	$590	
Humidity meter—10 at $79 each	$790	
Temperature probe—10 at $49 each	$490	
pH sensor—10 at $109 each	$1,090	
Batteries for handheld data collection tool	$50	
Digital camera	$400	
Laptop computer	$1,700	
Shipping and handling equipment	$60	
Lumber for cover boards—25 sheets at $20/sheet	$500	
Spring scales	$75	
Flagging tape, paint for cover boards, plastic baggies, rulers	$65	
PROFESSIONAL DEVELOPMENT OR TRAINING		Project director has no experience with digital sensors. A summer training institute will assist the project director in becoming familiar with the sensors and software. Funds will cover institute costs, transportation, lodging, and meals.
Four-day digital sensor summer institute/training	$1,600	
OTHER (MISCELLANEOUS SUPPLIES, TRANSPORTATION, ETC.)		The text will provide accurate protocols for amphibian monitoring used by field biologists. Salamander models will be used by students prior to actual field work to gain experience in how to accurately take snout-vent measurements.
Text: Measuring and monitoring biological diversity, standard methods for amphibians	$40	
Models of salamanders for measurement practice	$50	
STIPEND(S)		The project director will receive a modest stipend to cover the cost of time spent preparing lesson plans, establishing a study site, and attending the summer institute. Curriculum related to amphibian identification, use of sensors, and protocols related to salamander monitoring will be created.
Project director stipend—20 hours at $25/hour	$500	
TOTAL	$9,500	

to incorporate the curriculum writing as an in-kind service and to have instead budgeted money for transportation to field sites or for paying substitute teachers. Note that although the bulk of the proposal was related to hardware, it was essential for carrying out the grant proposal objectives. Also note that professional development training was incorporated into the proposed grant. Although the students did not participate in the training, it was integral to student learning.

Example Budget for a Small Grant (Less Than $1,000)

Some funding agencies awarding smaller grants might not require an extensive budget and justification. For example, the funded *Let's Go Solar* project had a very simple budget and no required justification, as shown in Table 4.4. Nevertheless, all items were related to the project and were listed, along with the name of the vendor and shipping costs.

TABLE 4.4
Example of a Budget for a Small Grant

ITEM	VENDOR	QUANTITY	INDIVIDUAL COST	TOTAL COST
SOLAR HOUSE	Carolina Biological	5	$54.80	$274.00
SOLAR POWER KIT	Carolina Biological	5	$14.50	$72.50
SOLAR ELECTRICITY DEMO	Carolina Biological	5	$14.45	$72.25
SOLAR CAR ASSEMBLY	Carolina Biological	4	$45.50	$182.00
SHIPPING COSTS	10% of order			$60.00
TOTAL				$660.75

Important Points to Remember

- Strictly follow the budget guidelines of the funding agency when preparing your proposal's budget.
- Make sure all purchases in the budget can be funded according to the granting organization's funding guidelines.
- Write the grant for the entire amount if possible.
- Only include items germane to your project's objectives.
- Address all major purchases with a justification that is closely tied to the proposed project's objectives.
- Include in-kind services or resources available from your school.

Exercises

1. Will you be purchasing equipment? If so, make a list of vendors you can contact if you have questions regarding the equipment.

2. Using the large grant budget example, determine where the author did the following:

   - Grouped small items into a broad category
   - Defended the purchase of major pieces of equipment
   - Included shipping and handling in the budget
   - Limited the budget to a specific investigation (e.g., habitat preference of *Plethodon cinereus*)
   - Justified the use of funds for professional development and a stipend

Your Turn!

Create your budget using the "Estimating the Budget" template in Appendix 1 (p. 109). List all major equipment and supplies, professional development or training, stipends, miscellaneous supplies, travel, and any other items that may not easily fall into one of the categories.

Evaluation: Insight Into Your Project's Success

The purpose of evaluation is to assess the overall success of your proposal and to determine if the project's goals and objectives have been met. As you develop your project goals, think about how best to measure and evaluate your level of success at meeting each objective. Incorporate evaluation early on in your project, documenting evidence of student learning on a continuous basis. Although it is not necessary to provide the actual instrument that will be used to evaluate your project's outcomes, having a structured plan in place from the beginning will make composing any reports that the funder stipulates easier to complete.

Any type of formative or summative assessment is acceptable when evaluating the project's success. Excellent methods for conducting formative assessment of students' learning and their level of engagement on an ongoing basis include journal reflections, photos, interviews, observations, and pre- and postsurveys designed to show changes in attitude. Record quotes from students as you hear them, and document student work with a camera or smartphone throughout the project so that you will be able to draw from a rich resource when you are writing the final report (see the Reporting to the Funder section in Chapter 8, pp. 83–85). If your objectives include specific science content or concepts that your students will learn as a result of their participation, consider administering a pretest and a posttest to gauge mastery or growth. Lab reports, presentations, and projects can also be used to gauge the impact the project will have on your students.

If the project involves partners and colleagues, consider what information you will collect from them to illustrate their contribution to the success of the project. For those projects that feature a level of community involvement, be sure to state how many people will be affected, their level of involvement, and how you will assess the project's impact.

Important Points to Remember

- Align your evaluation tools to your objectives.
- Use a variety of evaluation approaches (surveys, presentations, interviews, etc.) to provide evidence for the success of your project.
- Document evidence of student learning as you conduct your project.
- If applicable, include the community or any partners when developing your evaluation.

Evaluation Example

The example below, from a funded grant, includes a description of how the students were to be assessed. The purpose of the assessment was twofold: to measure student learning and to evaluate if the project objectives were met.

> *Students will be assessed using a variety of techniques, including essay responses, journaling, fieldwork, and data analysis. Students will demonstrate knowledge of field protocols, and will work cooperatively to develop stream assessment reports. Students will be able to assess water quality using a Biotic Index and chemical tests, and will demonstrate understanding of biological diversity and its dependence upon healthy water. Students will be able to describe how human use of the land has affected the quality of the stream.*

Exercises

1. Look over the goals and objectives for your grant proposal. Which one(s) would best be assessed using formative assessment? Which one(s) would best be measured using summative assessment? What qualitative and quantitative data can you collect that will illustrate that your objectives have been met?

2. Think about the type of questions you could include in a survey designed to measure change in attitude toward science.

Your Turn!

Use the "Writing the Evaluation" template in Appendix 1 (p. 110) to identify the evaluation methods you will use to demonstrate that your proposal's objectives have been met.

Dissemination Plan: Sharing Your Work With Others

As any science teacher knows, one of the most important steps in the scientific method is communicating your results so that others can learn about your work. Disseminating what you have accomplished as a result of your successfully implemented grant proposal is no exception. This is your opportunity to share the lessons you have learned with others. When creating a dissemination plan, begin by determining the audience with whom you would like to share your work, such as any of the following:

- Teachers in your school, your school district, or your science department
- Students in your school
- Teachers outside of your district
- Your students' parents
- Your school board
- Your local community
- Your state representative

There are several factors that will come into play when deciding on an audience. Did the work involve a community partnership? If so, consider having

your students present their findings or work to your city planning commission, the township board of supervisors, the local parks and recreation department, a gardening club, or any entity that would benefit from the information. Examine Figure 4.4; notice that it is suggested that *your students* present the project's key outcomes. Students find it extremely motivating to have an authentic audience with which to share the project outcomes. Consider providing your students with this opportunity, even if you teach at the elementary level.

Figure 4.4

Engage Students in Disseminating Project Results.

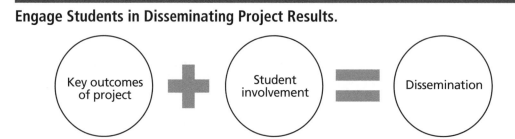

Within your school community, your students can make presentations to other students, the school faculty, or your district's school board. An alternative to making presentations would be to have your students create a public service announcement or a video, or to participate in a science expo event in which parents, teachers, and administrators have a chance to learn from the students.

Other options for engaging your students in communicating what they have learned include writing a press release, arranging for an article to be published in the local newspaper or school district website, or participating in an interview with your local television or radio station. Regardless of the venue, student participation in the dissemination process can create a positive relationship between the school and the community.

You may have noticed that your dissemination plan can be tied to your project's evaluation, because when you have the community participating in dissemination events, you will have a chance to evaluate how people in the community have been affected by your project. You will also be able to assess your students' ability to effectively communicate their project findings and learning experiences.

Disseminating to Colleagues

Perhaps your dissemination plan is simply to share what you have done and learned with your school colleagues at a faculty meeting or department meeting. You may want to consider broadening this audience to a state or national one. Obtaining a grant and carrying out a proposal places you in an elite group of teachers who have demonstrated their success and innovation outside of their classrooms. You are in the fortunate position of being able to effect change in other science classrooms that may be geographically far removed from you! Presenting at a county, state, or national science education conference is a great opportunity to disseminate the results of your project. Many states have active science teacher associations that sponsor an annual conference filled with sessions designed to bring K–12 teachers of science up-to-date on topics related to assessment, standards, methods, and pedagogy. NSTA regional or national conferences present additional options for dissemination. Challenge yourself to submit a session proposal; the particular venue you choose may depend on your personal comfort level, proximity, and funding. If your proposal is accepted, talk with your administration about your potential attendance at the conference. Although many school budgets are tight, some administrations are eager to support a teacher's attendance at a conference when the teacher is a presenter.

If your school or personal budget doesn't allow for you to present at an in-person conference, consider presenting at an online conference. Online conferences can be a lot of fun to attend as both presenter and attendee. It is thrilling to be able to interact with teachers from all parts of the United States and beyond. Regardless of what you decide, prepare for your dissemination by collecting photos of students engaged in carrying out the grant project, and take notes as you go along so that you will have the materials you need to create a great presentation. Be sure to obtain parental permission to take and use photographs of students in presentations.

Your dissemination plan can also consist of blogging, using Twitter, or participating in Twitter chats. For maximum impact, consider disseminating your project through NSTA's many social media outlets such as the NSTA Learning Center, Facebook, blogs, and e-mail lists. You may also want to consider writing an article for one of NSTA's K–12 journals. NSTA journals include a journal for elementary teachers, *Science and Children*; a journal for middle school science teachers, *Science Scope*; and a high school science teacher journal, *The Science Teacher*. If you plan to write an article, it is prudent to plan ahead, as it may take you longer than you anticipate to compose and submit an article for possible publication. You should be aware that NSTA requires a model release form to print any photographs of students. As such, it is wise to obtain signatures on the Model Release Forms as you carry out your project. If you have any questions concerning Model Release Forms or the appropriateness of your proposed article, you can contact the field editor of the appropriate NSTA journal.

Important Points to Remember

- Dissemination can be local, state, or national.
- If possible, engage students in the dissemination process.
- NSTA has many outlets for dissemination, including blogs, journals, and conferences.

Sample Dissemination Plan

The dissemination plan below, which was created for a grant proposal involving a professional development opportunity, makes use of the author's experience presenting at conferences.

On my return, I will make a presentation to my school board and science department and will author an article for our school district's community newsletter. Hands-on activities developed as a result of attendance at the Earthwatch Expedition will be offered via a K–12 inservice workshop during the summer of 2015.

As an active member of local, state, and national science teacher associations, I will conduct a number of outreach activities to share the knowledge acquired through my attendance at the Earthwatch Expedition. These activities will include presentations made at the local Science Teacher Association Conference in October 2015 and the state Science Teacher Association Conference in December 2015.

Exercises

1. Dissemination plans can take many forms. What are you comfortable doing? What dissemination activities are outside of your comfort zone?

2. How can you engage your students in a dissemination plan that is appropriate for their age and abilities?

3. Define how you can you use your dissemination plan to challenge yourself to grow professionally.

Your Turn!

Using the "Creating a Dissemination Plan" template in Appendix 1 (p. 111), describe your plan for disseminating your project idea.

Sustainability Plan

When writing your grant proposal, it is important to think about how you will sustain the project in the future. Although some funders will provide resources for a onetime event such as a field trip, many organizations prefer to provide seed money for a project whose life extends beyond one year. Some granting organizations will ask you to explain how you plan to sustain the project once the funding has expired. You will therefore need to give some thought to project sustainability if you plan to engage your students in the project activities annually. If your budget consists primarily of nonconsumable items such as computers or probeware, sustainability will not be an issue. If, however, the grant is funding items such as field trips, stipends, and consumable supplies, you will need to explain how you plan to sustain your project. You should talk with your administration to determine if the district will support your project's continuation. Fundraisers, your school's parent-teacher organization, and local business support, as well as charging for services or supplies, are various ways to raise money to sustain your project for the future.

Important Points to Remember

- Be prepared to explain how your project will continue into the future.
- Identify alternate sources of funding for the purpose of sustainability.

> **Exercise**
> Talk to your principal about alternative pathways that you can use to sustain your project in future years.

> **Your Turn!**
> Using the "Determining Project Sustainability" template in Appendix 1 (p. 112), describe your plan for project sustainability.

Tackling the Project Summary

The project summary should always be written last. Although you may be anxious to complete and submit your proposal by the time you get to this point, it is crucial for you to create a clear and concise summary of your grant proposal. The project summary can make or break your chances of getting funded because it is often the first impression grant reviewers have of your grant. When writing the project summary, be cognizant of its importance and the potential impact it has on the likelihood of your proposal's funding. A well-written summary pulls the readers in and heightens their interest, whereas a poorly worded one can leave the readers confused and unimpressed. If you are accomplished at writing concisely (and many science teachers are), then composing the summary will pose no problem for you.

When you are ready to write your summary, read over your grant proposal, pulling out the following key information:

- Why is this work important? (*needs statement*)
- Who is your target audience? (*population description*)
- What are your learning goal(s)? (*objectives*)
- Where will this work occur? (*project description*)
- What will be accomplished? Mention any partnerships or organizations with whom you will be working and their role (*methods*)
- How will you know you reached your goal? (*evaluation*)
- How will you share what you learned with others? (*dissemination plan*)

Compose your summary as if it were a movie trailer. Trailers hit the high points of a movie, giving you enough information to understand the story, but leaving you wanting more. Like a trailer, your summary should provide an overview for your reader but leave them with wanting to read the rest of the proposal. Hook the reader by starting with a sentence that will capture their interest; this is best done by focusing on what your students will be doing and learning rather than by starting with the identified need. Since a description of the need and target population will be addressed in depth in separate proposal components, it is recommended that you not spend precious words describing why your school district is needy. If your school is a Title I school or has a high percentage of English language learners or underserved students, you may want to include a few statistics to best depict the educational setting in which you teach.

> **Tip!** You have a lot of information to cover when writing the summary. Make sure that you add transitions to move the reader along from one topic to the next. Transition words help to connect ideas to each other. They are particularly important when trying to cover so much information in a short amount of space.

Be aware, however, that there are many schools filled with underserved populations. What will set your project apart from the rest is not information about need, but rather a well-written and concise summary that presents attainable objectives that are tied to student learning.

Important Points to Remember
- Reread your proposal, selecting the most important points.
- Write concisely.
- Think of your summary as a movie trailer.
- Focus on student learning rather than need.

Example of a Project Summary
The following summary statement was taken from a funded grant proposal. Notice how the opening statement draws the reader in by identifying a real-world problem. A brief discussion of the methods, student learning objectives, and evaluation follows. Note that it is not necessary to discuss the budget in the proposal summary. Keep the summary to no more than one page, paying attention to the requirements of the granting agency.

> *The ability to grow high-quality pesticide-free food in a restricted amount of space will be vital to the availability of our food supply in the future. This program will provide students with hands-on learning experiences in growing vegetable*

seedlings and herbs using a continuous-flow hydroponic system. The growing system will allow a team of 140 seventh-grade students to grow plants to maturity throughout the school year. It will also allow students to start seedlings in the classroom that will be used for an already existing on-site vegetable garden.

Student learning experiences will include monitoring the growing solution for nutrient content and pH values and recording growth of different plant species. Students will use Excel software to construct graphs of the data and will analyze the graphs for trends in plant growth. Students will also monitor the gardening system for the presence of insect and fungus pests, will identify any pests found, and will apply integrated pest management techniques to manage pests found within the gardening system. Students will also be challenged to create their own soilless system using recyclable materials such as plastic containers, glass bottles, or milk cartons. Data collected from the growth of plants using their soilless system will be compared to the continuous-flow system.

Students will be assessed using a variety of parameters including student-designed lab experiments, Excel-generated graphs, and understanding of lab protocols used to analyze pH and nutrient values.

Exercises

1. Why is the summary the most important part of the grant proposal?

2. Explain why the summary may be the most difficult aspect of the grant proposal to write.

3. Read over the example project summary. Locate the following components: needs statement, target population, objectives, project description, methods, evaluation.

4. Pick one of your lesson plans and try to summarize it in a page by including the goals and objectives, activities assessment and evaluation, and the student population and their needs in terms of learning science concepts and skills.

Your Turn!

Using the "Writing the Summary" template in Appendix 1 (p. 113), write a concise summary of your project. Give your summary to someone unfamiliar with your project and ask that person to assess it for clarity.

Supplemental Grant Components

CHAPTER 5

Depending on the funding organization that you are working with, they may require you to address or submit additional components such as a résumé, teaching philosophy, letters of support, and lesson plans. These components provide grant proposal readers with supporting information that serves to provide additional insight into your proposal and your abilities to successfully complete the proposed work.

Project Staff Résumé or Vitae

Potential funders want to know that you are capable of carrying out your grant proposal, which is the reason many organizations require you and any colleagues with whom you are working to submit a vitae or résumé. If you don't have a recent résumé, now is the perfect time to bring it up to date in terms of your education and

accomplishments. It is a good idea to update your résumé annually so that you are prepared to act quickly on grant applications and professional development opportunities should a possibility arise.

If it has been a while since you prepared a résumé, you may want to use the internet or your word-processing software to locate templates. When preparing your updated résumé, include your college degrees, teaching certifications, and participation in professional workshops and trainings. When summarizing your work experience, include your classroom teaching experience and any duties or activities that speak to your work as a professional educator. These could include curriculum writing, committee work, coaching, mentoring, and other activities that demonstrate your leadership ability outside your classroom. It is crucial that you also include any publications you have written, workshops you have presented, or blogs you maintain as evidence of your professionalism and willingness to contribute to the learning of other teachers. Remember to highlight any innovative teaching programs in which you have participated, as well as other funded grant proposals and anything that is relevant to your proposed project. Although a strong résumé is not essential to getting funded, it does communicate to the granting organization that you are capable of following through.

Important Points to Remember

- Remember that your résumé or vitae is evidence of your ability to successfully carry out a project.
- Update your résumé annually.
- Use your résumé to focus on your skills and experiences.

> **Exercise**
>
> Examine your most recent résumé. Do you think that a potential funder would feel satisfied regarding your ability to carry out a project? Can you identify areas that need to be updated, polished, or expanded?

> ## Your Turn!
>
> Using the "Items to Include on Your Résumé" checklist in the Appendix 1 (p. 115), create a list of all the items that you will include in your updated résumé or vitae.

Teaching Philosophy

Some grants and professional development opportunities require you to submit a teaching philosophy as part of the application. A teaching philosophy is an explanation of your approach to teaching. You probably have a teaching philosophy, even if you have never really thought about it or heard it referred to in this manner. Having a well-thought-out teaching philosophy prepared will allow you to take advantage of grants and professional development opportunities that cross your path with a very short turnaround time.

Writing a teaching philosophy requires you to reflect on your teaching and explain how your teaching mirrors your beliefs about how students learn. It is always a good idea to give specific examples so that the reader can better understand your classroom environment.

You may want to think about describing the following:

- Your role in the classroom (are you a facilitator of learning or a deliverer of information?)
- Your assessment practices and what you consider to be critical evidence of learning
- Methods you employ to ensure that all learners can access material
- How you communicate your goals to students
- How you ensure all students are learning

You may find it helpful to look over some sample teaching philosophies, such as the one in the next section.

Sample Teaching Philosophy

> *Analyzing and reflecting on my teaching to improve classroom instruction and achievement is a constant practice and is accomplished through self-analysis, articulation with colleagues, and discourse between my students and me. Although I teach the same subject all day, my classroom composition differs*

from class to class, necessitating self-analysis for each class of the day. Frequent formative assessments, observations, and discussions with small groups of students are conducted with the purpose of implementing immediate changes to improve student understanding of concepts and tasks. When my students are engaged, can articulate and demonstrate their understanding of connections between concepts, and are correctly completing tasks using higher-order thinking skills, I know that the lesson has been effective.

I have cotaught a section of special education students for the last seven years with a special education teacher. Daily articulation with my colleague regarding lesson adaptations and modification has led to increased student achievement in science. I have found that many of the adaptations and changes to instruction that I make for the special education students are often beneficial for regular education students as well. Recently, when introducing the concept of cells, instead of having the students read the chapter on their own, I read one paragraph at a time, stressing proper pronunciation of cell organelles' names. The students then summarized in one sentence what they perceived to be the main idea of the paragraph. Popsicle sticks were used to randomly select a student who read his or her answer. A brief discussion followed, during which the students modified their summaries as necessary. This strategy was so successful that I repeated it in subsequent regular education classes. At the completion of each class, I surveyed the students for their opinion regarding the activity; nearly all said that they enjoyed the activity and that it helped them better comprehend the reading passage. This type of discourse is useful because it both increases student participation and allows me to experience the lesson from a student's point of view, better enabling me to make modifications that increase student understanding.

Exercises

1. What would you add to or delete from the sample teaching philosophy?

2. Reflect on your own teaching. How do you believe students learn best and how does your classroom mirror that belief?

Your Turn!

Use the "My Teaching Philosophy" template in Appendix 1 (p. 116) to begin to draft your own teaching philosophy.

Letters of Support

Once you have written your grant proposal, share it with your principal or head-master and ask that person to write you a strong letter of support that shows that he or she read the proposal. The letter should convey enthusiasm for the project and the opportunity that will be available to the students should the project be funded. A sentence or two that expresses an understanding of how the students and the school will benefit will demonstrate to the funder that your principal fully understands the project and its potential for broadening student learning.

All letters of support should be written on letterhead, signed, and dated. Anything less can interfere with your grant being approved. In the past, letters of support were mailed along with the grant proposal. Today, many funders have moved to an online application process. Some may require you to scan in the letter, while others may provide a link for administrators to use to enter their letters directly. Still others may require your administrator to simply check a box denoting his or her approval. Regardless of the option, it is a good idea to request the letter of support from your administrator at least two weeks before you intend to submit your grant, since they may take a while to be generated.

You will also need a letter of support from any government entity, community organization, or university you are collaborating with. Make sure that the letter clearly delineates their role in your proposal and expresses their support and enthusiasm.

Sample Letter of Support

To Whom It May Concern,
December 12, 2012

I am writing this letter of recommendation on behalf of Ms. Smith, who is applying for funding through the grant program your foundation is administering. Ms. Smith has my full support to carry out the work on the project, The Effect of Materials on Windmill Design and Efficiency. *To support Ms. Smith's project, I have allocated $1,000 to cover the cost of substitute teachers, materials, and transportation for field trips, which meets the required (50%) for matching funds.*

When Ms. Smith came to discuss her project with me, she had already secured the support of our local university's engineering department. After reading the proposal, I was very excited that such project would take a place in our Title I school because our students will learn how to design and test their own windmills. Needless to say, having the support of engineers will give our students the opportunity to discuss their ideas with professionals.

I believe that Ms. Smith is fully capable of carrying out this unique project. She has been a sixth-grade science teacher at our school for the last five years. I have visited Ms. Smith's classroom numerous times, and I am always amazed with the vibrant classroom environment she creates for her students. Ms. Smith's impact extends beyond her classroom to encompass the entire student body. Last year, Ms. Smith was successful in securing funds to organize Innovation Night to showcase our students' science, engineering, and art projects. Innovation Night required a tremendous amount of work on Ms. Smith's part, but she was able to excite the faculty and enlist their support.

Ms. Smith has my full and unconditional support for her proposed project, The Effect of Materials on Windmill Design and Efficiency. *Please do not hesitate to contact me if you have any questions.*

Sincerely,

Dr. John Baker, Principal

Important Points to Remember

Check to see that the letter

- strongly supports your proposal;
- is signed and dated;
- is on school or organization letterhead;
- conveys that the person read the grant proposal (Does it mention the project and specify student learning outcomes?); and
- conveys enthusiasm for the proposal.

> **Your Turn!**
> Ask for your letter(s) of support at least two weeks prior to the grant deadline.

Lesson Plans

You may be asked to provide a sample lesson plan to complete your grant application. For example, the National Science Teachers Association (NSTA) Shell Science Lab Challenge application requires teachers to submit a description of an activity that they currently teach using minimal resources. Naturally, you will want your lesson to be well connected to your proposal's learning objectives. Strive to include an innovative and exemplary lesson that engages the students; a verification lab activity or a worksheet that came with your textbook is neither exemplary nor innovative. As mentioned in the Project Description (Methods) section (p. 33–37), you should engage your students in authentic data collection that is age appropriate and aligned with the funder's mission. The purpose of asking for a lesson plan is for you to demonstrate your ability to identify a learning task and have your students engaged in carrying out that task. It is also a wonderful opportunity for you to help the grant reader visualize what learning looks like in your classroom.

Some grants will not require a lesson plan but will give you the option to submit one. Always submit any optional documents because this will separate your proposal from those that do not. Anything you can do to make your proposal more professional and complete will help your chances of receiving the grant. If the grant does not ask for a lesson plan, do not include one, since it is critical that you follow the exact requirements of the grant application.

Important Points to Remember

- Be prepared to provide a lesson with your grant proposal if required or suggested by the funding agency.
- Make sure the lesson is tied to your grant objectives.

Exercise

Review your favorite lessons and select two or three that could serve as examples of your teaching.

CHAPTER
6

NGSS: A Valuable Tool for Designing Winning Grant Proposals

The Next Generation Science Standards *take the position that a scientifically literate person understands and is able to apply core ideas in each of the major science disciplines, and that they gain experience in the practices of science and engineering and in crosscutting concepts. (NGSS Lead States 2013, p. xxiii)*

The statement above summarizes the ideas behind the design of the *Next Generation Science Standards* (*NGSS*) and the importance of developing a scientifically literate population (the *NGSS* can be downloaded for free from the National Academies Press at *http://www.nap.edu/catalog/18290/next-generation-science-standards-for-states-by-states*). The standards themselves are arranged by both disciplinary core idea and by topic; either format can be used as a resource when writing your grant proposal. Using the *NGSS* to develop your grant proposal is beneficial because you will be aligning your goals and objectives with standards that reflect the most current research in science education. Even if your state has yet to adopt or will not be adopting the *NGSS*, you will find the standards to be an excellent resource to use when writing your grant proposal. By using the *NGSS* as a blueprint, you can ensure that your proposal will be well aligned with engineering and science practices, connected to other science disciplines, and linked to *Common Core State Standards* (*CCSS*; NGAC and CCSSO 2010; see Figure 6.1).

Elements of the *NGSS* align strongly with several of the components typically included in grant proposals, such developing your project's goals and objectives, identifying methods you will employ throughout your project, justifying the needs of your target population, writing the project summary, and developing your assessments and evaluation tools. After using the *NGSS* to outline these proposal components, you will only have the timeline, the budget, and the evaluation to complete—components that can be easily developed after you have the foundation of your project written. Now that you are already familiar with the grant proposal components, let's look at how you can use the *NGSS* in the grant proposal writing process.

Figure 6.1

Using the *NGSS* to Draft Components of a Grant Proposal

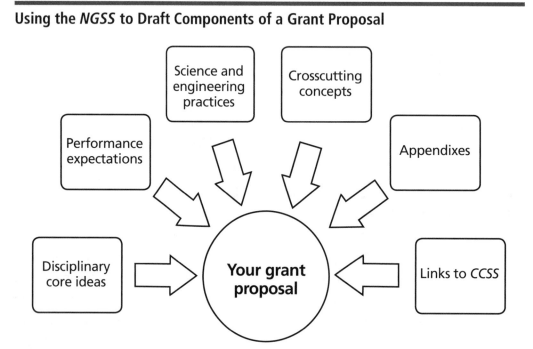

Using the *NGSS* Disciplinary Core Ideas as a Source of Project Ideas

As previously discussed in Chapter 3 (p. 15), your district and state standards can act as a source of ideas for a potential classroom project. Similarly, an examination of the *NGSS* disciplinary core ideas for the grade level you teach can reveal a wealth of ideas that can serve as inspiration for a project. Disciplinary core ideas and subideas describe what students should know and understand about essential concepts of the major science disciplines (*NGSS* Lead States 2013). To help you better visualize how the *NGSS* can be used in the grant proposal writing process, this chapter will use *Let's Go Solar*, a proposal that involves middle school students in learning about solar power as an alternative energy resource. The inspiration for *Let's Go Solar* came from the *NGSS* standard MS-PS3 Energy (see Figure 6.2, pp. 68–69).

To get started, examine the following project summary for *Let's Go Solar*:

> *This project is designed to teach sixth- and seventh-grade gifted students about the benefits of using solar power as an alternative energy resource and the importance of developing alternative energy technologies for the future. The students will use models of solar cars, solar houses, solar-powered lightbulbs, and fans to determine how solar energy is transformed into thermal, mechanical, and electrical energy. The project will expand students' understanding of the importance of new energy resources and why working to create alternative energy technologies will help us meet the increased demand for energy in the world. The students will present an evidence-based argument for the question, "Why doesn't South Florida depend more on its available abundant solar power?" This project will engage students in the scientific and engineering practices of planning and carrying out an investigation, analyzing and interpreting data, constructing explanations and arguments, and engaging in argument from evidence.*

The U.S. dependency on nonrenewable energy sources is grounded in the following disciplinary core ideas for middle school, which helps to identify the concepts the students will learn.

PS3.A: Definitions of Energy

Temperature is a measure of the average kinetic energy of particles of matter. The relationship between the temperature and total energy of a system depends on the types, states, and amounts of matter present.

Figure 6.2

MS-PS3 Energy Served as a Source of Inspiration for *Let's Go Solar.*

MS-PS3 Energy

MS-PS3 Energy
Students who demonstrate understanding can:

MS-PS3-1. **Construct and interpret graphical displays of data to describe the relationships of kinetic energy to the mass of an object and to the speed of an object.** [Clarification Statement: Emphasis is on descriptive relationships between kinetic energy and mass separately from kinetic energy and speed. Examples could include riding a bicycle at different speeds, rolling different sizes of rocks downhill, and getting hit by a wiffle ball versus a tennis ball.]

MS-PS3-2. **Develop a model to describe that when the arrangement of objects interacting at a distance changes, different amounts of potential energy are stored in the system.** [Clarification Statement: Emphasis is on relative amounts of potential energy, not on calculations of potential energy. Examples of objects within systems interacting at varying distances could include: the Earth and either a roller coaster cart at varying positions on a hill or objects at varying heights on shelves, changing the direction/orientation of a magnet, and a balloon with static electrical charge being brought closer to a classmate's hair. Examples of models could include representations, diagrams, pictures, and written descriptions of systems.] [Assessment Boundary: Assessment is limited to two objects and electric, magnetic, and gravitational interactions.]

MS-PS3-3. **Apply scientific principles to design, construct, and test a device that either minimizes or maximizes thermal energy transfer.*** [Clarification Statement: Examples of devices could include an insulated box, a solar cooker, and a Styrofoam cup.] [Assessment Boundary: Assessment does not include calculating the total amount of thermal energy transferred.]

MS-PS3-4. **Plan an investigation to determine the relationships among the energy transferred, the type of matter, the mass, and the change in the average kinetic energy of the particles as measured by the temperature of the sample.** [Clarification Statement: Examples of experiments could include comparing final water temperatures after different masses of ice melted in the same volume of water with the same initial temperature, the temperature change of samples of different materials with the same mass as they cool or heat in the environment, or the same material with different masses when a specific amount of energy is added.] [Assessment Boundary: Assessment does not include calculating the total amount of thermal energy transferred.]

MS-PS3-5. **Construct, use, and present arguments to support the claim that when the kinetic energy of an object changes, energy is transferred to or from the object.** [Clarification Statement: Examples of empirical evidence used in arguments could include an inventory or other representation of the energy before and after the transfer in the form of temperature changes or motion of object.] [Assessment Boundary: Assessment does not include calculations of energy.]

The performance expectations above were developed using the following elements from the NRC document *A Framework for K-12 Science Education*:

Science and Engineering Practices	Disciplinary Core Ideas	Crosscutting Concepts
Developing and Using Models Modeling in 6–8 builds on K–5 and progresses to developing, using and revising models to describe, test, and predict more abstract phenomena and design systems. • Develop a model to describe unobservable mechanisms. (MS-PS3-2) **Planning and Carrying Out Investigations** Planning and carrying out investigations to answer questions or test solutions to problems in 6–8 builds on K–5 experiences and progresses to include investigations that use multiple variables and provide evidence to support explanations or design solutions. • Plan an investigation individually and collaboratively, and in the design: identify independent and dependent variables and controls, what tools are needed to do the gathering, how measurements will be recorded, and how many data are needed to support a claim. (MS-PS3-4) **Analyzing and Interpreting Data** Analyzing data in 6–8 builds on K–5 and progresses to extending quantitative analysis to investigations, distinguishing between correlation and causation, and basic statistical techniques of data and error analysis. • Construct and interpret graphical displays of data to identify linear and nonlinear relationships. (MS-PS3-1) **Constructing Explanations and Designing Solutions** Constructing explanations and designing solutions in 6–8 builds on K–5 experiences and progresses to include constructing explanations and designing solutions supported by multiple sources of evidence consistent with scientific ideas, principles, and theories. • Apply scientific ideas or principles to design, construct, and test a design of an object, tool, process or system. (MS-PS3-3) **Engaging in Argument from Evidence** Engaging in argument from evidence in 6–8 builds on K–5 experiences and progresses to constructing a convincing argument that supports or refutes claims for either explanations or solutions about the natural and designed worlds. • Construct, use, and present oral and written arguments supported by empirical evidence and scientific reasoning to support or refute an explanation or a model for a phenomenon. (MS-PS3-5) - ***Connections to Nature of Science*** **Scientific Knowledge is Based on Empirical Evidence** • Science knowledge is based upon logical and conceptual connections between evidence and explanations (MS-PS3-4),(MS-PS3-5)	**PS3.A: Definitions of Energy** • Motion energy is properly called kinetic energy; it is proportional to the mass of the moving object and grows with the square of its speed. (MS-PS3-1) • A system of objects may also contain stored (potential) energy, depending on their relative positions. (MS-PS3-2) • Temperature is a measure of the average kinetic energy of particles of matter. The relationship between the temperature and the total energy of a system depends on the types, states, and amounts of matter present. (MS-PS3-3),(MS-PS3-4) **PS3.B: Conservation of Energy and Energy Transfer** • When the motion energy of an object changes, there is inevitably some other change in energy at the same time. (MS-PS3-5) • The amount of energy transfer needed to change the temperature of a matter sample by a given amount depends on the nature of the matter, the size of the sample, and the environment. (MS-PS3-4) • Energy is spontaneously transferred out of hotter regions or objects and into colder ones. (MS-PS3-3) **PS3.C: Relationship Between Energy and Forces** • When two objects interact, each one exerts a force on the other that can cause energy to be transferred to or from the object. (MS-PS3-2) **ETS1.A: Defining and Delimiting an Engineering Problem** • The more precisely a design task's criteria and constraints can be defined, the more likely it is that the designed solution will be successful. Specification of constraints includes consideration of scientific principles and other relevant knowledge that is likely to limit possible solutions. *(secondary to MS-PS3-3)* **ETS1.B: Developing Possible Solutions** • A solution needs to be tested, and then modified on the basis of the test results in order to improve it. There are systematic processes for evaluating solutions with respect to how well they meet criteria and constraints of a problem. *(secondary to MS-PS3-3)*	**Scale, Proportion, and Quantity** • Proportional relationships (e.g. speed as the ratio of distance traveled to time taken) among different types of quantities provide information about the magnitude of properties and processes. (MS-PS3-1),(MS-PS3-4) **Systems and System Models** • Models can be used to represent systems and their interactions – such as inputs, processes, and outputs – and energy and matter flows within systems. (MS-PS3-2) **Energy and Matter** • Energy may take different forms (e.g. energy in fields, thermal energy, energy of motion). (MS-PS3-5) • The transfer of energy can be tracked as energy flows through a designed or natural system. (MS-PS3-3)

Connections to other DCIs in this grade-band: **MS.PS1.A** (MS-PS3-4); **MS.PS1.B** (MS-PS3-3); **MS.PS2.A** (MS-PS3-1),(MS-PS3-4),(MS-PS3-5); **MS.ESS2.A** (MS-PS3-3); **MS.ESS2.C** (MS-PS3-3),(MS-PS3-4); **MS.ESS2.D** (MS-PS3-3),(MS-PS3-4); **MS.ESS3.D** (MS-PS3-4)
Articulation across grade-bands: **4.PS3.B** (MS-PS3-1),(MS-PS3-3); **4.PS3.C** (MS-PS3-4),(MS-PS3-5); **HS.PS1.B** (MS-PS3-4); **HS.PS2.B** (MS-PS3-2); **HS.PS3.A** (MS-PS3-1),(MS-PS3-4),(MS-PS3-5); **HS.PS3.B** (MS-PS3-1),(MS-PS3-2),(MS-PS3-3),(MS-PS3-4),(MS-PS3-5); **HS.PS3.C** (MS-PS3-2)
Common Core State Standards Connections:

*The performance expectations marked with an asterisk integrate traditional science content with engineering through a Practice or Disciplinary Core Idea.
The section entitled "Disciplinary Core Ideas" is reproduced verbatim from A Framework for K-12 Science Education: Practices, Cross-Cutting Concepts, and Core Ideas. Integrated and reprinted with permission from the National Academy of Sciences.

May 2013   ©2013 Achieve, Inc. All rights reserved.   1 of 2

Figure 6.2 (*continued*)

MS-PS3 Energy Served as a Source of Inspiration for *Let's Go Solar.*

MS-PS3 Energy

ELA/Literacy –	
RST.6-8.1	Cite specific textual evidence to support analysis of science and technical texts, attending to the precise details of explanations or descriptions *(MS-PS3-1),(MS-PS3-5)*
RST.6-8.3	Follow precisely a multistep procedure when carrying out experiments, taking measurements, or performing technical tasks. *(MS-PS3-3),*(MS-PS3-4)
RST.6-8.7	Integrate quantitative or technical information expressed in words in a text with a version of that information expressed visually (e.g., in a flowchart, diagram, model, graph, or table). (MS-PS3-1)
WHST.6-8.1	Write arguments focused on discipline content. *(MS-PS3-5)*
WHST.6-8.7	Conduct short research projects to answer a question (including a self-generated question), drawing on several sources and generating additional related, focused questions that allow for multiple avenues of exploration. (MS-PS3-3),*(MS-PS3-4)*
SL.8.5	Integrate multimedia and visual displays into presentations to clarify information, strengthen claims and evidence, and add interest. *(MS-PS3-2)*
Mathematics –	
MP.2	Reason abstractly and quantitatively. (MS-PS3-1),(MS-PS3-4),(MS-PS3-5)
6.RP.A.1	Understand the concept of ratio and use ratio language to describe a ratio relationship between two quantities. (MS-PS3-1),*(MS-PS3-5)*
6.RP.A.2	Understand the concept of a unit rate a/b associated with a ratio a:b with b ≠ 0, and use rate language in the context of a ratio relationship. *(MS-PS3-1)*
7.RP.A.2	Recognize and represent proportional relationships between quantities. (MS-PS3-1),*(MS-PS3-5)*
8.EE.A.1	Know and apply the properties of integer exponents to generate equivalent numerical expressions. (MS-PS3-1)
8.EE.A.2	Use square root and cube root symbols to represent solutions to equations of the form $x2 = p$ and $x3 = p$, where p is a positive rational number. Evaluate square roots of small perfect squares and cube roots of small perfect cubes. Know that $\sqrt{2}$ is irrational. *(MS-PS3-1)*
8.F.A.3	Interpret the equation $y = mx + b$ as defining a linear function, whose graph is a straight line; give examples of functions that are not linear. (MS-PS3-1),*(MS-PS3-5)*
6.SP.B.5	Summarize numerical data sets in relation to their context. *(MS-PS3-4)*

Source: NGSS Lead States 2013.

ESS3.C: Human Impacts on Earth Systems

Typically, as human populations and per-capita consumption of natural resources increases, so do the negative impacts on Earth, unless the activities and technologies involved are engineered otherwise.

ETS1.B: Developing Possible Solutions

- A solution needs to be tested, and then modified on the basis of the test results, in order to improve it. (MS-ETS1-4)

- Models of all kinds are important for testing solutions. (MS-ETS1-4)

ETS1.C: Optimizing the Design Solution

- Although one design may not perform the best across all tests, identifying the characteristics of the design that performed the best in each test can provide useful information for the redesign process. (MS-ETS1-3)

- The iterative process of testing the most promising solutions and modifying what is proposed on the basis of the test results leads to greater refinement and ultimately to an optimal solution. (MS-ETS1-4)

Exercise

Read over the *NGSS* disciplinary core ideas for the grade level and/or disciplinary content you teach. Identify the disciplinary core idea(s) that you could use as the basis for your grant proposal idea.

Develop Your Project Goals Using the *NGSS* Performance Expectations

Once you have an idea for a grant proposal, you can begin developing goals and objectives using performance expectations from the *NGSS*. The *NGSS* define performance expectations as "the assessable statements of what students should know and be able to do" (NGSS Lead States 2013, p. xxiii) Performance expectations encourage development of critical thinking skills, communication, and problem-solving skills, as well as science and engineering practices in the teaching and learning of science. The inclusion of *NGSS* performance expectations in your proposal will help your students achieve high standards and increase the value of your proposal.

The inspiration for *Let's Go Solar* project goals can be found in the following *NGSS* performance expectations for the topic of energy at the middle school level.

MS-PS3-3: Energy
Apply scientific principles to design, construct, and test a device that either minimizes or maximizes thermal energy transfer.

MS-ESS3-3: Earth and Human Activity
Apply scientific principles to design a method for monitoring and minimizing a human impact on the environment.

MS-ETS1-3: Engineering Design
Analyze data from tests to determine similarities and differences among several design solutions to identify the best characteristics of each that can be combined into a new solution to better meet the criteria for success.

MS-ETS1-4: Engineering Design
Develop a model to generate data for iterative testing and modification of a proposed object, tool, or process such that an optimal design can be achieved.

These performance expectations were then rewritten to a specific goals that defined what students would be doing as part of the *Let's Go Solar* proposal. Students were expected to do the following:

- Plan and carry out an investigation to determine how solar energy is transferred into heat, mechanical, and electrical energies.

- Plan and carry out an investigation using solar cars as a model to determine the conditions for their best performance. In this process, identify independent and dependent variables and controls, what tools are needed to do the data collection, how measurements will be recorded, and the quality and quantity of data needed to support a claim.

> ## Exercise
>
> Read over the *NGSS* performance expectations for the grade level and/or disciplinary content you teach. Think about what you would add to justify the need for the *Let's Go Solar* project or come up with your own idea for a project based on these performance expectations.

Outline Your Project Activities With the *NGSS* Science and Engineering Practices

Each performance expectation found in the *NGSS* is supplemented with science and engineering practices that support its learning goals. These practices are intended to engage students in the type of work conducted by scientists and engineers. Within the practices associated with a specific performance objective are statements that you may be able to use to develop your project methods. In the case of *Let's Go Solar,* students are engaged in the following science and engineering practices:

- Constructing explanations and designing solutions
- Developing and using models
- Planning and carrying out investigations
- Analyzing and interpreting data
- Engaging in argument from evidence
- Obtaining, evaluating, and communicating information

Using the science and engineering practices associated with the previously identified performance expectations, the following specific activities were developed for *Let's Go Solar:*

- Plan and carry out an investigation, individually and collaboratively, using solar cars as a model to determine the conditions for their best performance. In this process, identify independent and dependent

variables and controls, what tools are needed to do the gathering, how measurements will be recorded, and the type and quantity of data needed to support a claim.

- Plan and carry out an investigation to determine how solar energy is transferred into heat, mechanical, and electrical energies.

- Construct, use, and present oral and written arguments supported by empirical evidence and scientific reasoning to support or refute the use of solar power as a source of energy.

- Construct an argument supported by evidence for how the use of solar energy along with developing alternative energy technologies could decrease the impact of the increased consumption of natural resources on Earth's systems.

Exercise
What science and engineering practices support your grant proposal idea?

Develop an Integrated Science Approach Using the Crosscutting Concepts

The *NGSS* performance expectations are also supported with crosscutting concepts that are used to explain phenomena in the natural world across different science disciplines.

Crosscutting concepts can be used to connect your project to other disciplinary core ideas at your grade level and can help you to identify performance expectations that could be taught in conjunction with one another. The inclusion of crosscutting concepts will add depth to your project by connecting your project's objectives to other science disciplines and will enhance your students' understanding of the natural world. Given that many of the issues facing the world today will require an integrated science approach, crosscutting concepts can provide a source of inspiration for connecting your classroom work to a real-world experience for your students.

Let's Go Solar addresses the following crosscutting concepts related to cause and effect and energy and matter (linkages to performance expectations are found in parentheses):

Cause and Effect

Relationships can be classified as causal or correlational, and correlation does not necessarily imply causation. (MS-ESS3-3)

Energy and Matter

- Within a natural or designed system, the transfer of energy drives the motion and/or cycling of matter. (MS-ESS2-4)
- The transfer of energy can be tracked as energy flows through a designed or natural system. (MS-PS3-3)

Systems and System Models

Models can be used to represent systems and their interactions—such as inputs, processes, and outputs—and energy and matter flows within systems.

Exercise

How can you connect the performance expectations you have identified for your proposal idea to a real-world setting?

Supporting Your Target Audience's Needs With the *NGSS*

The *NGSS* are intended for all students; use them to show how your target population will benefit from the project you are proposing. For example, English language learners (ELLs) would benefit from hands-on exploration of science concepts because they are learning science and improving their language skills when working and discussing within their teams. You will find a great deal of supporting information in the *NGSS* appendixes that you can use when writing your grant proposal. In particular, Appendix C, which addresses college and career readiness, and Appendix D, which emphasizes that the standards are written for all students, can be invaluable at helping you to justify your identified need while making science accessible to all students. Examination of the *NGSS* appendixes may help you identify many additional ideas that will serve to benefit the students you teach.

Examine the target population for *Let's Go Solar*. Notice how the author linked career and college readiness to meeting the needs of gifted learners.

> *Middle school students are of an impressionable age and are passionate about making this world a better place. In a few short years, they will be either entering college or selecting a career. Engaging gifted learners in an authentic task may*

stimulate students to pursue a future STEM [science, technology, engineering, and mathematics] career. Additionally, the multidisciplinary approach of physical science combined with engineering will help students to better understand the connections that naturally exist been disciplines.

Exercise

What would you add that could be useful for regular, gifted, or advanced students; ELLs; or students with various disabilities?

Develop Interdisciplinary Projects Using *NGSS* Linkages to the *CCSS*

The *NGSS* also provide linkages to the *CCSS ELA* (English language arts) and literacy and the *CCSS Mathematics* (NGAC and CCSSO 2010). Including these interdisciplinary connections will indicate that your project examines the specific science topic through the lenses of different disciplines. When it comes to developing grant proposals of an interdisciplinary nature, elementary teachers have a decided advantage since they can easily integrate reading, writing, and mathematics into science. Middle and high school teachers can strengthen their proposals through collaboration with English, mathematics, and social studies teachers to design projects that will have a main science theme applicable to the concepts covered in these subjects' classes. Interdisciplinary projects can also involve collaborations with physical education, art, and music teachers, which could bring a different dimension to your project and tap into your students' creativity. No matter what grade level you teach, clearly depicting the connections between science, mathematics, and English language arts will enhance the planned learning experiences for your students and make your proposal more appealing to the grant reviewers.

In the case of *Let's Go Solar*, the following *CCSS ELA*/literacy standards were addressed:

RST.6-8.1

Cite specific textual evidence to support analysis of science and technical texts, attending to the precise details of explanations or descriptions.

RST.6-8.3

Follow precisely a multistep procedure when carrying out experiments, taking measurements, or performing technical tasks.

WHST.6-8.1

Write arguments focused on discipline-specific content.

> **Tip!** No matter what grade level you teach, requiring students to keep science notebooks will assist you in easily making connections between science and English language arts.

WHST.6-8.7

Conduct short research projects to answer a question (including a self-generated question), drawing on several sources and generating additional related, focused questions that allow for multiple avenues of exploration.

> **Exercise**
>
> Use the *NGSS* to find connections to *CCSS ELA*/literacy standards and *CCSS Mathematics* standards that you can embed into your project.

Develop Your Project's Assessments Using the *NGSS*

As stated earlier, the *NGSS* considers performance expectations to be "assessable statements of what students should know and be able to do" (NGSS Lead States 2013).

Note how assessment for *Let's Go Solar* has been developed directly from the science and engineering practices.

> *There will be formative and summative assessment of the students' progress. In addition, students will perform self-assessment throughout the project. Students will be assessed on their ability to plan and carry out the investigations individually or collaboratively, to gather and analyze data, and construct explanations of their experimental findings. The students will receive individual grades on their lab reports of the performed investigations. Class discussions of the films* An Inconvenient Truth *and* Who Killed the Electric Car? *will require students to present their arguments and support them with evidence. Arguments will be assessed to evaluate student understanding of the concepts of alternative energy technologies and their potential to remediate the negative effect of excessive burning of fossil fuels.*

Exercise

How can the science and engineering practices that you identified as being germane to your grant proposal idea be developed into assessments?

Your Turn!

Begin filling out the "Using the *NGSS* as a Blueprint" worksheet in Appendix 3 (p. 121).

References

National Governors Association Center for Best Practices and Council of Chief State School Officers (NGAC and CCSSO). 2010. *Common core state standards.* Washington, DC: NGAC and CCSSO.

NGSS Lead States. 2013. *Next Generation Science Standards: For states, by states.* Washington, DC: National Academies Press. *www.nextgenscience.org/next-generation-science-standards.*

CHAPTER 7

Submitting Your Grant: Keep Those Fingers Crossed!

In all likelihood, you will be submitting your grant online. Depending on how the system is set up, you may be able to start a grant application, save it, and come back to it later. If it does not allow you to save your work, you will have to enter all the information at once. Regardless of the setup, it may be helpful to compose your responses to the grant proposal sections in a word-processing document. This will allow you to easily spell-check your work and will save you frustration in the event that your computer malfunctions or you lose internet access while completing the online application. If there is a word or character count limit, make sure that you adhere to it, as the online submission form will usually automatically cut off anything that runs over the limit. Gather all of your documentation prior to beginning the actual online application process. Have an electronic copy of all necessary documents ahead of time, including letters of reference, lesson plans, proof of tax-exempt status, résumés, or any other material required by the granting organization. It is strongly recommended that you keep a copy of all documents in case your computer crashes or some other disaster befalls you. Once you are satisfied with the quality of your proposal and have gathered all the required paperwork in an easily retrievable manner, begin uploading your materials into the website. Prior to hitting the "submit" button, double check to make sure that you have carefully copied all demographic data and proposal sections into the submission form and that all required files and documentation items have been uploaded.

> **Tip!** Prior to submitting your grant proposal, have someone proofread and evaluate it using the "Grant Proposal Rubric" found in Appendix 2, pp. 117–119.

If the funding organization requires you to mail your proposal, bear in mind that grant reviewers are more concerned with the actual proposal rather than a fancy binder or other presentation tools. If you are mailing in the grant, follow all the requirements precisely. For example, if the organization specifies a maximum number of pages to be stapled in the upper left corner, prepare and submit your grant with only that number of pages and be sure the staple is in the correct corner. Following directions communicates to the funder your ability to carry out your proposal successfully.

Important Points to Remember

- Write your grant using a word-processing document.
- Double-check for grammar and spelling issues.
- Stay within the word or character count limit.
- Gather all the necessary files ahead of time and have them available electronically.

Exercise

Gather all of the documents you need and store them in electronic form both on your hard drive and in the cloud.

> **Your Turn!**
> Identify colleagues who could be helpful in proofreading and evaluating your grant proposal.

Dealing With Rejection

Despite your best efforts, there is always the possibility that your grant proposal will not be funded. It is important to not get disheartened should this happen, as even successful grant proposal writers are familiar with having their ideas rejected. It can be very difficult to receive that rejection notification when you have worked hard on an idea that you are passionate about implementing. There are a number of reasons that can cause a proposal to not getting funded. It could be that there were many excellent applications for limited funds, that your idea did not meet the funder's mission and objectives, or that your vision was not clearly and concisely communicated to the grant reviewers. Although you have no control over your competition, you can potentially increase the probability of receiving funding by following the ideas outlined in this book.

Like the process of writing a grant proposal, being confronted with an unfavorable result is a learning experience. One positive action you can take is to call the grant manager and ask for specific feedback regarding your proposal. Once you have received the feedback, you have one of three options available: You can (1) revise and resubmit the grant to the same funding organization, (2) submit your grant proposal to a different organization, or (3) come up with a new idea altogether.

Tips! Although these next few tips won't guarantee success, they will help increase your chances of receiving funding.

- Carefully look over the funding organization's website to ensure that your idea aligns with its mission.
- Read over the types of grants that the organization has funded in the past. Is your idea something that seems to fit with what it generally funds?
- Most granting organizations have a grant manager or program officer on staff. Call the grant manager and run your idea past him or her to gauge if it is well-matched to the organization's mission.
- Ask the grant manager if it is possible for you to get a copy of a previously funded grant.
- Ask the grant manager if he or she would look over your grant prior to submission.

You can increase your understanding of the review process by volunteering to be a grant reviewer for an organization that funds proposals. In many cases, the people who review grant proposals for organizations do not possess a teaching background and would welcome your input. By reading grant proposals, you can quickly get a sense of the quality of work that makes some proposals shine above the rest. Whatever you decide to do, it is important to remember that risk taking is linked to success and that you have already distinguished yourself as a teacher leader by being willing to take the time to pursue an opportunity that will benefit student learning.

Important Points to Remember

- Grant proposal writing is a skill, but there are many things you can do to increase your chances of being funded.
- Writing is hard work, but the effort is worth it!
- Call the grant manager to seek his or her input.
- If rejected, call the grant manager to see how you can make improvements.
- Revise and resubmit.

Managing Your Funded Project

CHAPTER
8

You've identified a dream, written a grant proposal, and received notification either via e-mail or letter that your proposal will be funded. Now begins another exciting period—making your dream a reality. It is during this time that you will engage your students in science and engineering practices and challenge their minds. This will be a very rewarding experience for you, for your students, and for any of your colleagues involved in the project. Please keep the suggestions in this chapter in mind as you proceed with carrying out activities related to your grant.

Managing Your Funded Project

The first bit of paperwork related to receiving the grant involves writing a brief handwritten thank-you note to the funding agency. This is an instance in which a typed letter or e-mail will not suffice; a handwritten note is a nice gesture to show your appreciation to the funding agency for giving you the chance to make your dream a reality. It is a good idea to send the thank-you note to the program manager as soon as you are notified that you have been awarded the grant.

Your next item of business should be to map out your strategy for carrying out your proposal. Start by referring back to your timeline. Now that you have received the grant, you will need to add in the date(s) that your grant report is due. If your plan for dissemination included presenting at a conference, start the conference proposal process. Stay on track with your project by entering the timeline into the calendar on your phone, printing it out and placing it on the wall in your office, or using any technique you have found that helps you to stay organized. As your project unfolds, remember to document its success via pictures and notes.

Spending the Project Funds

If you received a letter, it may have contained a contract, a check made out to your school, or both. The first thing you should do (after jumping up and down with glee) is to inform your supervisor of your success. If your notification included a contract to be signed by the business manager or school treasurer, principal, and/or superintendent, ask your principal's assistance in ensuring that the contract is signed and returned expeditiously.

If your notification letter contained a check, your supervisor will probably forward the check to the business office or to the school treasurer where a special account will be set up for you (obtain the account number from your business office so that you have it in the event that the business manager or the school secretary should leave your district). Although this is a school account and the funding agency gave the money to the school, you will need to make it clear that the money is dedicated to supporting your project and is not to be used for activities or supplies unrelated to the grant.

Once you are ready to order equipment or supplies, you will need to have a conversation with your administrator to determine the preferred method of ordering, since accounting regulations will often stipulate how the account is to be established and who has access to it. Some schools may require that the school secretary or treasurer make all purchases, while other schools will reimburse you if you purchase items on your own. Prior to purchasing items, obtain a copy of your school's tax-exempt form from the principal's office or your school district's business office. If you show a copy of the form at checkout, you will not have to pay tax. Regardless of the protocol, you will need to work within your school district's accounting procedures.

You should keep a record of all purchases and make sure that the balance reflects the balance in your grant account. This will make your life a lot easier when you start to write the required progress reports or the final report for the funded project. You should decide how to keep your records—hard copy or electronic—but good record keeping is essential. Sometimes schools get audited, and some of your purchases might be questioned for the simple reason that the person auditing the school may not understand the nature and the goals of your project. Having meticulous records will be of great value to the school should such a situation arise.

Whether purchasing locally or through a vendor, you will need to determine if using a school purchase order is an option (many items, especially consumable items, can be far cheaper when purchased locally). If you have a preferred vendor that is not on the school vendor list, check with your school business manager about getting the vendor approved. To obtain the best prices, you may want to investigate online vendors; just make sure that you incorporate shipping and handling into the price.

You will need to contact the grant administrator with a sound justification if you need to make changes to your purchasing plan. Valid reasons for changing the planned purchases include finding better deals for your supplies or adding additional activities to the project. Most grant administrators will try to accommodate any reasonable and well-justified request for changes in the budget to help ensure the successful completion of the project.

Reporting to the Funder

As part of your obligation to the funder, you will be expected to provide some sort of report that summarizes your progress. Each funding organization has specific requirements for reports that may be driven by the duration of the project and the amount of money awarded to you. You may be expected to provide a single report at the conclusion of the project, or you may be responsible for submitting one or two progress reports in addition to an end-of-project report. Some reports will need to answer certain questions, while others will require you to describe in detail how the project is progressing and how the funds have been spent but with no specific questions to answer.

When writing the report, you will usually be asked to describe your project's progress on your goals, the number of people who benefited and a description of the benefit, and the reaction from the community regarding your project. You may also be asked to provide an accounting of all actual expenditures compared with your proposed budget. You should be prepared to offer a detailed explanation about the state of the budget following the requirements of the funding agency. Most funders will also ask you to provide photographs, newspaper clippings, or links to websites that feature information related to your project. You will find it helpful to prepare for these reports by collecting documentation throughout the life span of your project.

In your report to the funding agency, you should clearly describe what project activities have taken place and how those activities have contributed to the progress of the project. The other very important component of these reports is explaining what your students have learned as a result of working on the project and how you have assessed their learning. Consider these reports a showcase of your efforts and validation of your dedication to your students' learning, so include as many details as possible. Always include comments about how your students have reacted to the project, how they have been affected, and any other anecdotes that you find appropriate. If you have worked with a college or university, a state or national park, or businesses, include a description of the partnership in the progress report even if these contacts are in an early stage. This will show that you have accessed community resources for the benefit of your students.

It is important for you to understand that the purpose of the report is not to pass judgment on your ability to carry out your proposal (in fact, when writing progress report you will have a chance to inform funding agencies about any problems you encountered or changes to the project that you would like to initiate). Although the report helps to inform the funder about the successes and challenges of your project, the funder is aware that projects don't always turn out as planned. The funder's purpose in asking for a progress report is to assess the grant program and how it can better serve the community. You will not be penalized if your project is somewhat behind on its timeline or does not meet all of its objectives. Rather, the funder is genuinely interested in your project and its impact on your students and community.

Although writing progress reports is your obligation to the funding agency, it may be helpful to know that you are not the only one creating reports; program managers are also authoring similar reports. They use the information submitted by grantees to discuss the effect the funding program has on students and teachers and K–12 science education. Many of these programs raise their money from different sources, and the future of the grant program financing depends on the performance of the current grantees. It is therefore important to do the best job possible when writing your report and to submit all reports on time. To ensure that your report is submitted on time, you may want add the report due dates to your timeline and calendar and set up a reminder on your smartphone a week before the report is due.

Important Points to Remember

- Grant reports are required by the funder to assess the success of the funding.

- You will be expected to write a final report and possibly several progress reports.

- Collect evidence as you conduct your proposal.

- Submit your reports on time!

Some Advice as Your Project Unfolds
Continuously Collect Data About Your Project

A very important task while working on your project is collecting evidence about your project's progress and how your students are affected by their participation. Do the following to ensure that you have the as much accurate data as possible:

> **Tip!** With careful observation, you could have enough data to write an article or make a presentation.

- Write down anything your students say in reference to the project. Make it a habit to do that at the end of each day.

- Observe students carefully while they are working and record your observations. How has you project affected the overall student learning that takes place in your classroom?

- Connect your class assessments to your project.

Engage in Good Public Relations

An activity you should engage in while working on your project is making sure people in school and the community know what you are doing. One place to start is the daily school news. Another way to spread the word about your project is by organizing presentations around the school where the participating students can discuss their work with their peers. You can also ask your students to write a short story or article about the project and publish it in the school newspaper.

These public relations activities can also extend outside of school. Try contacting your local newspaper or a television station; these entities enjoy showcasing successful stories like yours. With reports about how U.S. students are falling behind international students in subjects like math and science, it is refreshing to hear how our students are engaged in authentic scientific investigations. You may want to consider making a presentation about your project at local or state conferences. Keep in mind that nothing is more inspiring than sharing your experience with other teachers who are genuinely interested in student learning.

Initiate, Nurture, and Establish Relationships

The importance of establishing relationships with businesses and educational entities in your community was discussed in Chapter 2 (p. 7). By continuing to seek partnerships throughout the course of your project, you will create new opportunities for yourself and your students. Do not be afraid to ask researchers from local universities to visit your classroom. If you live in a rural area and there is no university in close proximity, make use of technology to bring in scientists virtually for a video chat.

Enlist Help!

Being responsible for carrying out project activities in addition to your daily workload can be stressful, making it vital to secure all the help you can. Your best helpers are your students; most likely, they will enjoy taking part in the managerial aspect of the project. The list of jobs that the students can do is endless: organizing materials, being responsible for the cleaning and maintenance of the science lab, typing lab directions, making copies, writing articles, and making presentations to other students. Designating duties related to your project will benefit both you and your students; tasks will be accomplished, and your students will learn new skills.

Collaborate with colleagues. You undoubtedly work with many outstanding colleagues who are eager to work with you. Fortunately, science can be easily connected to any subject, making collaborating with a grade-level team or teachers of other disciplines a relatively simple venture. Such a mutually beneficial situation will help your students see the interconnectedness between subjects and will be a rewarding one for you and your colleagues. In addition, collaboration can be a source of new ideas for future projects.

Project Sustainability

The purpose of project sustainability is to make a lasting impact on more than one group of students. While engaged with your project, be on the lookout for opportunities to continue your project in the coming years by securing assistance from other sources. Such sources include local businesses, which may help you with supplies; graduate students or professors, who can share their expertise; or your own district, which can allocate funds to continue your project in the future. Keep in mind that you will need to learn how to identify the opportunities around you and not be shy about asking for help. Sustainability does not necessarily mean the continuation of the same project; sometimes, it means that you can use the results of your project to start a new one on a related topic. In that case, always make sure that you describe how the new project is connected to the original project and explain how this continuity ensures its sustainability.

Hiccups With the Grant

You should be aware that it is unlikely that your project will be completed exactly as you envisioned. Be prepared for the unexpected difficulties that may make carrying out your proposal challenging. In any situation, draw on your enthusiasm and passion for the project, your problem-solving skills, and your obviously incredible communication skills. Whatever unfolds, remember to keep your administrator up to date, and if necessary, consult with the grant administrator as well. With your skills and with some assistance, you will be able to navigate successfully through any challenge, whether it is a piece of equipment that is no longer available or a change in your teaching assignment. Below are some scenarios that you may encounter as you carry out your project.

I have a new principal who doesn't seem supportive of the project. You've already been tremendously successful at selling your idea to the funder, so there is no reason that you cannot be successful doing the same thing with your principal. Make an appointment to discuss the project and prepare for the meeting by documenting how your students will be learning. You should be able to explain why your methods are better than traditional methods. You can support your arguments by finding research that supports the pedagogical approaches occurring in your classroom. Don't forget to invite your principal to drop by any time to see the project in progress!

The equipment I planned to purchase and use is unavailable. Unless the equipment is extremely specialized, you should be able to work around this problem. First, call the vendor to see if they have any extra equipment in inventory. Just because they no longer appear to carry it doesn't mean that their inventory is completely devoid of the product. You can also call the manufacturer to see if they have any more of the product or if they have a new product that they recommend.

I found a different piece of equipment than the one I proposed buying. If the equipment purchase makes up only a small percent of your budget (under 10%) you should be able to purchase it without too much trouble. If you are purchasing something totally new (perhaps instead of the single computer you budgeted for you would prefer to buy several iPads) you should call the grant manager to make sure that he or she will approve the purchase. Remember that the grant was given to you with the contingency that the project will be carried out as outlined in the proposal. As long as the equipment substitution will help meet the proposal's objectives and is being used in the same manner as the originally budgeted equipment, the grant manager is likely to approve the change to your budget. You will, of course, have to clear the changes with your administration as well. This will be especially true if you, your administrator, or the business manager signed a contract stipulating that the grant money would be spent as outlined in the proposal.

My teaching assignment has changed. Depending on the specific scenario, this can be a tricky challenge. Perhaps you have been reassigned to a new grade level, changed schools, or are no longer in the classroom due to a change in your personal situation or a promotion to an administrative position. Depending on the circumstances, you have several options. If you have changed grade levels, you may be able to modify the project to make it suitable for a younger or older audience, or you may be able to convince a colleague to carry out the proposal as originally written.

If you have changed schools within the district, you might want to consider pursuing completing your project at your new school. This will mean meeting with both of your administrators (new and old) to ensure that equipment purchases are transferred to your new setting. Depending on the nature and perceived value of the equipment, your old principal may be reluctant to agree to this. For example, he or she may prefer that any technology equipment earned from the grant remain at your former school. If this is the case, you should let the two administrators figure out a solution. If you explain your project goals in a convincing manner to your new administrator, he or she will be more likely to assist you.

If you have moved to a new school district, you will not be able to take any equipment with you, since the grant money and items purchased with it belong to your previous school district. You will need to inform the granting organization that you will be unable to carry out the project as proposed. The grant manager will take care of the situation at that point.

Things are not working out as I had planned. This is actually a fairly typical scenario. Teaching situations change, things take longer than anticipated, and unforeseen complications arise. Your first steps are to identify the problem and possible solutions; be sure to involve your administrator in helping you to brainstorm for solutions. Next, call the grant manager and explain the situation. You'll find that most grant managers are calm and understanding when it comes to unexpected problems with carrying out proposals. Since grant managers have a great deal of experience administering grants and are more than likely knowledgeable of similar situations, they can be a tremendous resource in helping you brainstorm solutions.

I am having problems making the grant report deadline. You promised to submit a report about your project by May 1, but there is no possible way for your project to be completed by that date. Call the grant manager and let him or her know of the issue. Usually, the grant manager will give you an extension to allow you to finish the project as proposed, although he or she may ask for a mid-progress report by May 1 and a final report due at a later date that has been mutually agreed on by you and the grant manager.

The district won't allow me to spend my money. Although it is understandable that administrators might view the grant money as something that they can access, it is important to remember that the granting agency provided funds for a specific purpose. Failure to use the money for the manner in which it was stipulated is considered a breach of contract, which is something that most districts try to avoid. Start by talking with your administrator and sharing your concerns. If the meeting

Tip! Take care of the project director—*you*! Being a classroom teacher is stressful enough without the added responsibility of managing a project and reporting the project's progress to the funding agency. Aside from the general suggestions of staying positive and organized, eating right, exercising, and resting, you may want to request release from a small portion of your teacher duties to be able to take care of the grant's day-to-day activities. For example, you may want to request to have your homeroom period dedicated to planning experiments and discussing the project results with students. Approach your school principal, and have a conversation with him or her about the benefits your students will receive as a result of implementing the project. It is a good idea to already have some of your colleagues collaborating with you to show that you will be able to reach more students, which will add to the power of your request. If you teach at the high school level, discuss with your administration the possibility of dedicating a portion of your teaching schedule to your project, in which students can conduct related experiments and receive credit for an elective science class.

does not produce the desired results, you may have to call the business manager or superintendent. As a final alternative, you can call the grant manager. Since your project is most likely one that will be completed within one school year, try to spend all of your funds within the fiscal year in which you received your grant or you may run the risk that the district will roll your money into a common account.

I need an extension after the deadline to finish the project due to school days lost as a result of snow days in our school district. What should I do? Estimate the effect of these lost days on your project and develop a new timeline to try to remediate the situation. As soon as you have this information, contact the funder, explain the reasons why your project will be delayed, propose the new deadline, and ask for a permission to proceed with the revised timeline.

I am trying to collaborate with other teachers in the school, but no one seems willing to commit. Don't let this discourage you; perhaps your peers don't feel they have enough information to make a commitment. Talk to your principal about allowing you to make a short presentation during a teacher planning day or faculty meeting so that you can explain what you are doing, how your students will benefit, and how you envision your collaboration with the teachers in the school. Demonstrate an upbeat and confident attitude as you convey your excitement with the idea of collaborating with them.

A local TV station that would like to produce a short segment about our project has contacted me. Discuss the offer with your administration, and if they give you permission, inform the TV station that it is welcome to do the segment but that you have to get permission from the parents of the participating students. After that, proceed with collecting the parents' permission forms to videotape their children. When all the forms have been collected, decide what you would like to convey regarding your project, and then call the station. Although this will be an

> **Tip!** Don't get overwhelmed. Carrying out a grant can be a big job, but remember that your project was selected over others because it was truly amazing and the granting organization believes in *you*! So relax and enjoy the ride!

interruption to your classroom activities, your students will feel special and you will have an opportunity to promote how students are learning via your project.

I have established a productive collaboration with a teacher in my school, and everything was going fine until now. My colleague is now asking to change the course of the funded project. Congratulations on establishing a successful collaboration. Start by saying that you are very happy about your work together, but remind your colleague that you have an obligation to the funding agency to carry out the project the way it was described in the original proposal. Listen to your colleague's ideas, identify possible funding agencies for his or her idea, and if it is mutually agreeable, work on a grant proposal for a future submission.

Exercises

1. Create a conversation scenario with your principal to ask for release time to work on your project.

2. Develop a plan to approach colleagues for collaboration.

3. Create a conversation scenario or an e-mail message that asks a professor or business owner for help with your project.

You've Only Just Begun!

CHAPTER 9

Learning how to develop a successful grant proposal is a skill that will be useful in opening the door to many more exciting opportunities that will benefit your students and help you grow as a professional educator. There is a wealth of choices available to you, including awards, fellowships, and educational travel. This chapter briefly introduces you to some of the possibilities that await.

Opportunities for Recognition

Awards

There are several organizations that recognize teachers who carry out innovative science projects in their classrooms and who can demonstrate their contribution to student learning. Award applications are greatly enhanced if they include projects funded on a competitive basis by outside agencies. For example, the National

Science Teachers Association (NSTA) gives several awards annually that include monetary rewards, travel support to attend the NSTA national conference, and funds to purchase supplies for the classroom. These awards enhance individual teacher professional development and also serve to increase the visibility of the particular school science program. NSTA is not the only association that presents awards that recognize excellence in teaching. At the national level, you can apply for the prestigious Presidential Award for Math and Science Teaching. Numerous teacher associations for specific science disciplines also have annual awards to recognize outstanding teachers of biotechnology, biology, chemistry, Earth science, agriculture education, technology, and others. Very often, your local and state science teachers associations also present teaching awards. At the local level, teachers and students who complete environmental projects and contribute to increased awareness of environmental issues are eligible for recognition from Sierra Club chapters. This list is not comprehensive, but serves to provide you with an idea of the opportunities available to you.

National Board Teaching Certification

Applying to become a National Board Certified Teacher requires a substantial amount of writing, but it is a thought-provoking and rewarding experience. One of the entries requires you to document your professional development endeavors and how they have affected your students. Having a successful grant proposal funding history will provide you with material and examples for explaining the impact that funded projects have had on student learning.

Personal Growth Opportunities
Fellowships

There are many fellowships that will require you to submit a well-written proposal as part of the application. One of them is the Fulbright Distinguished Award in Teaching. Successful applicants spend three to six months taking courses at one of the country's universities for the purpose of learning about innovative teaching practices in the selected country. If you enjoy experiencing adventurous educational travel, you may be a good candidate for this award, but you will need to obtain approval from your school district prior to applying. If you get selected for this program, you will also be able to apply for alumni grants.

Another opportunity is the Albert Einstein Distinguished Educator Fellowship, which provides science teachers with the opportunity to serve 11 months at federal agencies such as the U.S. Department of Energy, the National Science Foundation (NSF), the National Oceanic and Atmospheric Administration (NOAA), and the National Aeronotics and Space Administration (NASA). Although this fellowship requires relocation to Washington, DC, it presents an excellent venue for exploring the science education policy environment. A well-written application is a prerequisite for acceptance, and your newly acquired grant writing skills will be extremely helpful in this process.

Travel and Expeditions

There exist many opportunities for science teachers to travel and participate in exciting projects. The numerous benefits of attending professional development outside your school district include gaining fresh ideas to bring back to your classroom and developing collaborations with teachers from around the country. Many professional development venues come with travel, accommodation, stipends, and funds for classroom supplies. You'll find that completing an application for professional development is similar to writing a grant proposal, requiring you to list goals and explain how your travel will affect student learning. Following participation in these programs, it is typically expected that you will deliver workshops to increase fellow teachers' content knowledge and the visibility of the programs. Applicants who have garnered funded grant proposals and experience with giving workshops to other teachers are at an advantage because they already have experience with what is expected from them after the professional development program has taken place.

Some examples of professional development opportunities at the national level that are fully funded and available to teachers across the United States include such well-known programs as NOAA Teacher at Sea, PolarTREC (Teachers and Researchers Exploring and Collaborating), the U.S. Department of State exchange program, the FDA/NSTA Food Science Workshop, and the Maury Project. You may even want to consider writing a grant to fund your own unique opportunity, such as participation in a graduate course that occurs in the tropics, or working alongside a scientist through attendance at an Earthwatch Expedition. Regardless of the type of expedition you choose to seek, your grant writing experience will allow you to demonstrate how attendance through experiences outside the classroom will allow you to extend student learning.

Research Experience for Teachers

Teachers can also apply for a Research Experience for Teachers. This is a very interesting opportunity for a science teacher to become engaged in NSF-funded science research at a local university. NSF is very interested in reaching out to more K–12 teachers by exposing them to the research process that takes place in science research laboratories. If you would like to join a university researcher's team, the principal investigator on the grant will need to submit an application for a supplemental grant to accommodate you in the researcher's laboratory over the summer.

Classroom Opportunities

STEM Competitions

Whether or not you chose to incorporate STEM (science, technology, engineering, and mathematics) competitions into your classroom is up to you, but realize that your students' interest in science will grow as a result of their exposure to the work conducted in your classroom as part of your funded proposals. With mentoring and encouragement, your students may be very successful at competing in local science fairs and national student competitions. Some examples of national competitions are the Toshiba ExploraVision competition, eCybermission, the DuPont Challenge Science Essay Competition, and the Bright Schools Competition. Winning students are typically awarded savings bonds, money, or trips, and they may be recognized at special events. Needless to say, winning any of these competitions will increase the visibility of the STEM program in your school district.

Many higher education institutions organize summer research camps for junior and senior high school students. You can act as a great resource for your students by helping them prepare successful applications for summer research opportunities. For example, the U.S. Department of Energy has a high school internship program that places students to work on research projects in the department's National Laboratories across the United States. Students who are guided through the process by their teacher have an advantage in that they may be more likely to submit a successful application.

Share Your Knowledge With Others
Write an Article for a Science Education Journal

Working on your funded classroom project will yield valuable insight into pedagogy and student learning. One of the ways to share this knowledge with your colleagues is to write an article for submission to a practitioner journal, such as one of the NSTA journals. You may also want to consider contributing to a blog or to a newsletter. It cannot be stressed enough how important it is to have teachers publishing and sharing their innovative, instructional practices with other teachers. Becoming a published author brings enhanced credibility to any future grant proposal application because it demonstrates that you have used previous funding not only to leave a lasting impact on your students' learning but also to disseminate your newly acquired knowledge to your colleagues and other researchers. If you find writing articles to be stimulating and rewarding, it may even lead to writing a book or getting a research degree in science education.

Present at a Science Education Conference

Another venue to share your funded project experiences with other educators is through presentations at relevant local, state, and national conferences. NSTA hosts several annual conferences, but if you are thinking of submitting a proposal, make sure that you comply with the submission deadlines since conferences often receive more submissions than they can accommodate. Other organizations that sponsor teaching conferences include the National Association of Biology Teachers, the International Society for Technology in Education, the Association for Middle Level Educators, the National Association for the Education of Young Children, and the National Association for Gifted Children. This list is not exhaustive but provides you with many possibilities for outreach.

Host a Workshop

You may want to consider offering workshops for teachers in your district during which you present the instructional practices that you have developed as a result of implementing the projects in your classroom. Teachers often value proven methods that have been developed and tested by other classroom teachers, and working on funded projects with your students will provide you with anecdotes and insights that you can share with your colleagues.

Service Opportunities

Become a Grant Reviewer

Becoming successful at obtaining funding for your projects will make you a valuable member of the review teams of funding agencies. You are the practitioner who knows the types of grant ideas that can work in the classroom. Consider volunteering or applying for these reviewer positions by talking to the grant program manager and informing him or her that you would like to be considered as a reviewer in the future.

Serve on a Science Education Advisory Board or Committee

Each year, NSTA publishes the leadership positions available for NSTA review panels, committees, and advisory boards. Consider applying for those that you are most interested in and feel best suited for. Practicing teachers are in a unique position to make meaningful contributions because they have insight into K–12 classrooms and student learning. Alternately, you may want to consider becoming more involved with your state science teachers association, the Council for Elementary Science International, or the National Middle Level Science Teachers Association.

Exercises

1. Select two teacher awards, visit the award websites, and become familiar with the requirements to submit an application.

2. Visit your home state department of education and identify the grants available for districts and schools.

Some Final Words of Advice

CHAPTER 10

It is our hope that you will be fully supported by your administration, your colleagues, your students, and their parents as your funded project unfolds. Because your project may result in deviation from the "standard" curriculum, you may experience situations or conversations in which you will find yourself having to explain or possibly defend your choices. Be prepared for these opportunities to educate others about your project and its potential for influencing and improving student learning. Be confident in your abilities to carry out your proposal. The granting agency that funded you believes in you and in your power to create an environment that promotes student learning.

We recommend that you convey information about your project and goals to parents prior to beginning your project. Explain to parents that your students will benefit from participation in the project and describe the anticipated results for

their child's learning. If necessary, have research studies available that support your decisions. You'll find that most parents will be thrilled that their child has a teacher who is willing to plan and carry out engaging and meaningful science lessons. These parents will become your best advocates, touting the wonderful work you are doing to anyone that will listen. Parents represent an incredible resource; enlist them to help you to organize materials, make phone calls, or assist with hands-on activities. You'll find that parents who spend time in your classroom are driven to do so because they believe in you and in your approach to learning. The type of relationship that emerges between you and the parents of your students can be extremely rewarding as you work together to meet the needs of your students.

Continue to use your developing leadership skills to affect change within your building so that other students can have an opportunity to benefit from an innovative classroom project. You can do this by collaborating with fellow faculty members, providing training to your colleagues regarding the use of equipment you may have garnered through your grant writing, or by sharing new skills and pedagogy that you may have acquired. For those colleagues who are apprehensive when it comes to involving students in student-driven inquiry, offer to coach or mentor them. Not only will your passion for science spread throughout your grade level or school, but also your efforts will result in fostering a professional learning community that respects the strengths and talents of individuals for their unique contributions.

No doubt your dream to write a grant proposal has been inspired by the students who sit in your classroom. Your desire to improve student learning and bring innovation into your classroom will help you successfully carry out your proposal. Your students, however, are not the only ones who will benefit. In the process, you will learn valuable skills related to communication, problem solving, and communication. You will better understand the value of patience and the necessity of perseverance when pursuing goals and dreams. As part of your dissemination plan, you may challenge yourself to give a presentation at a meeting or to write an article for the local newspaper or a science educator journal. When being evaluated, your funded grant will provide evidence of your exemplary teaching ability. Long after the activities related to the grant have passed, we encourage you to continue to hone your skills as you grow into a teacher leader capable of effecting change outside of your own classroom. A whole world of teaching awards, student competitions, travel fellowships, and teacher researcher opportunities are within your reach. We encourage you to take advantage of it!

Dream Big! Grant Proposal Writing Templates

APPENDIX
1

DREAM BIG!

What is YOUR dream? What are YOU passionate about?

Describe your dream:

WRITING THE NEEDS STATEMENT

1. What problem or need exists in your school, school district, or community?

2. What evidence of need can you provide?

3. How will your proposal address this need?

4. Who will be affected by your project?

5. How will students benefit from your idea?

Check! Did you …

- ☐ identify the need?
- ☐ cite evidence in the form of statistics, National Science Teachers Association (NSTA) position statements, or research?
- ☐ explain how your proposal will meet the need?
- ☐ describe the target population?

DOCUMENTING TARGET POPULATION DEMOGRAPHICS

School District Information

- What is the percentage of students in your school who receive free or reduced-price lunch? _____
- What is the percentage of students in your school who are Title I? _____
- What is the population of your school? _____
- Type of school: _____ Public _____ Private
- School location: _____ Rural _____ Urban _____ Suburban

School Ethnicity

Write the approximate percentage (%) of students in each category. Enter whole numbers; the sum of all percentages should equal 100%.

_____ American Indian or Native Alaskan

_____ Asian

_____ Black or African American

_____ Native Hawaiian or other Pacific Islander

_____ Caucasian

_____ Mixed ethnicity

_____ Hispanic

Student Population

Write the approximate percentage (%) of students under each category.

_____ ELL students

_____ Special education students

Number of Teachers Affected

Total number of teachers grant will affect: _____

Number of Students Affected

Total number of students grant will affect: _____

Age or grade level of students: _____

What percentage of students affected fall into the following categories?

_____ Special education

_____ Regular education

_____ Gifted

Additional School District Information

_____ Mean income

_____ Mean state income

_____ Dropout rate

_____ Graduation rate

_____ State test scores and/or recent changes in scores

_____ Growth in any special populations

_____ Student-teacher ratio

_____ Student-computer ratio

Information About Your Town or City

_____ Mean income

_____ Mean state income

_____ Other

Check! Did you ...

☐ include information about your target population?

☐ include information that explains the need?

☐ use the National Center for Education Statistics to locate information related to your school?

☐ use up-to-date information?

WRITING YOUR GOALS AND OBJECTIVES

1. What are the learning goal (s) and/or learning objectives of your grant proposal?

2. How will your proposal help students to master concepts better than traditional teaching methods?

3. List all standards that will be met with your project.

Check! Did you ...

- ☐ create SUPER goals?
- ☐ write concisely?
- ☐ limit the number of goals?
- ☐ link your goals to standards?
- ☐ avoid educational jargon?

WRITING THE METHODS AND ACTIVITIES SECTION

1. What specific activities and investigations will your students be conducting?

2. Will you be working with other professionals and/or organizations? If so, what will be their roles?

Check! Did you ...

□ create activities that are directly related to the grant objectives and goals?

□ link your methods to student learning?

□ incorporate inquiry into your methods?

□ include partnerships?

□ involve an authentic audience?

□ limit the scope of your activities?

□ make sure all methods are age appropriate?

CREATING THE TIMELINE

MONTH	ACTIVITIES
AUGUST	
SEPTEMBER	
OCTOBER	
NOVEMBER	
DECEMBER	
JANUARY	
FEBRUARY	
MARCH	
APRIL	
MAY	
JUNE	

Check! Did you ...

☐ demonstrate that you have carefully thought out all aspects of your proposal?

☐ break down major events, such as trainings, purchases, pre- and posttests, and activities, and include the amount of time each event will take?

☐ relate each activity to an approximate date?

☐ add in extra time in case unexpected things occur?

ESTIMATING THE BUDGET

EXPENSE	ESTIMATED COST	JUSTIFICATION
MAJOR EQUIPMENT SUPPLIES		
PROFESSIONAL DEVELOPMENT		
STIPEND(S)		
IN-KIND SERVICES		
OTHER (TRANSPORTATION, ETC.)		

Check! Did you …

- ☐ record items and their estimated cost obtained from vendors?
- ☐ group items into major categories?
- ☐ write a justification for all big-ticket items?
- ☐ include in-kind services or resources available from your school?
- ☐ avoid padding the budget with unnecessary items?
- ☐ review the grant foundations restrictions on expenses to ensure that you are within their budget?

WRITING THE EVALUATION

1. What quantitative data will you submit to demonstrate that your proposal's objectives have been met? (Consider items such as pre- and posttests and attitude surveys.)

2. What qualitative data will you submit to demonstrate that your proposal's objectives have been met? (Consider items such as journals and portfolios.)

3. What is your plan for documenting ongoing evidence of student learning throughout the project? Did you remember to plan to obtain parental permission to take photographs of students?

Check! Did you ...
- align your evaluation to your objectives?
- create a plan for documenting student learning as you conduct your project?
- if applicable, include your community or partners in your evaluation plan?
- explain how you will document evidence of student learning throughout the project?

CREATING A DISSEMINATION PLAN

1. Identify who your audience will be. Consider disseminating information about your project to your state representative, local community, school board, science department, and/or teachers outside your district.

2. Will students play a role in disseminating? If yes, describe how.

3. Which methods do you plan to use to publicize your project?

   □ Press release
   □ News story in local paper
   □ News story on local television channel
   □ School district website
   □ Radio announcement

4. If applicable, which of the following methods will you employ for sharing with other teachers?

   □ Face-to-face conference
   □ Online conference
   □ Twitter
   □ Personal blog
   □ NSTA e-mail list or Learning Center Forum
   □ Article submitted to an NSTA journal

Check! Did you ...
   □ challenge yourself to go outside your comfort zone in planning for dissemination?
   □ involve your students in dissemination?
   □ consider a variety of ways to disseminate your project and/or share student learning?

DETERMINING PROJECT SUSTAINABILITY

1. Will the project continue in the future? Why or why not?

2. If yes, describe the support your district will provide for continuing the project in future years.

3. Identify alternate sources of funding for the purpose of sustainability, such as fundraisers, local business support, and charging for services or supplies.

Check! Did you ...

☐ explain how the project will continue in the future?

☐ list alternate sources of funding?

WRITING THE SUMMARY

1. Explain why the work is important by briefly describing the need (see the Describing the Need and Potential Impact section in Chapter 4, p. 25).

2. Briefly describe the most important demographics of your target audience (see the Describing the Target Population section in Chapter 4, p. 29).

3. State your learning goals (see the Developing Your Project Goals and Objectives section in Chapter 4, p. 31).

4. Describe where the work will occur (see the Project Description [Methods] section in Chapter 4, p. 33).

5. Briefly summarize the most important objectives that will be accomplished (see the Project Description [Methods] section in Chapter 4, p. 33). Mention any partnerships or organizations with whom you will be working with and their role.

6. Briefly summarize your plan for evaluating your project's success (see the Evaluation: Insight Into Your Project's Success section in Chapter 4, p. 47).

Check! Did you ...

- ☐ identify the need?
- ☐ define your target audience?
- ☐ discuss your learning goal(s)?
- ☐ explain where the work will occur?
- ☐ explain what will be accomplished?
- ☐ describe any partnerships or organizations with whom you will be working?
- ☐ describe your method for evaluation?
- ☐ limit the length according to your grant requirements?
- ☐ add transition elements?
- ☐ write clearly and concisely?
- ☐ ask a colleague to read it over for clarity?

ITEMS TO INCLUDE ON YOUR RÉSUMÉ

Contact Information

- ☐ Name
- ☐ Address
- ☐ E-mail address
- ☐ Phone number(s)

Education

- ☐ Degrees
- ☐ Certifications
- ☐ Badges
- ☐ GPA (only if above a 3.5 and you graduated in the previous two years)

Experience

- ☐ Teaching
- ☐ Coaching
- ☐ Summer employment (only if it demonstrates leadership)
- ☐ Youth leader (e.g., church, synagogue, Boy Scouts, or Girl Scouts) Involvement with innovative teaching programs

Professional Development

- ☐ Conferences attended
- ☐ Workshops or trainings

Teacher Leadership

- ☐ Presentations given
- ☐ Articles written
- ☐ Department work
- ☐ Mentoring student teachers

Other Evidence

- ☐ Previously funded grants
- ☐ Anything relevant to the proposed project
- ☐ Teaching awards
- ☐ Address of teacher website
- ☐ Military experience
- ☐ Volunteer work

Possible Skills to Include

- ☐ Fluency in another language
- ☐ CPR/first aid certification
- ☐ Computer skills

Check! Did you ...

- ☐ limit your résumé to two pages?
- ☐ include only the most recent and significant activities?
- ☐ avoid educational jargon?

MY TEACHING PHILOSOPHY

Answer the following prompts below to organize your thoughts when writing your teaching philosophy.

1. I believe students learn best by …

2. My role in the classroom can best be described as …

3. If someone were to observe my classroom, he or she would see …

4. I know my students are learning when …

5. If my students were to describe how I teach, they would say …

Grant Proposal Rubric

APPENDIX
2

ITEM	EXEMPLARY	ADEQUATE	NEEDS IMPROVEMENT
TARGET POPULATION AND DESCRIPTION OF NEED	The target population is clearly defined and includes any special education or minority populations. The description clearly explains the need and how the proposal will affect that need.	The target population is broadly defined; it may not include any special education or minority populations. The description adequately explains the need and how the proposal will affect that need.	The target population is weakly defined; there is no discussion of special education or minority populations. The need is not addressed.
POTENTIAL IMPACT	The impact clearly explains how student learning will be increased as a result of the proposed activities.	The impact adequately explains how student learning will be increased as a result of the proposed activities.	The impact weakly explains how student learning will be increased as a result of the proposed activities.
OBJECTIVE(S)	The objectives are clearly defined, linked to standards, attainable, and measurable.	The objectives are adequately defined, linked to standards, attainable, and measurable.	The objectives are vaguely defined, may or may not be linked to standards, may be unattainable, or are not measurable.
DESCRIPTION OF ACTIVITIES/ METHODS	All activities are directly tied to the grant proposal and are clearly defined, detailed, age appropriate, and realistic. A strong sample lesson plan is included (if required by the granting agency).	All activities are directly tied to the grant proposal and are adequately defined, detailed, age appropriate, and realistic. A sample lesson plan is included but may not be tied strongly to the proposed activities (if required by the granting agency).	The activities are weakly tied to the grant proposal. The activities may not be clearly defined, detailed, age appropriate, or realistic. The proposal may contain a weak or missing sample lesson plan (if required by the granting agency).
TIMELINE	The timeline demonstrates evidence of a well-thought-out proposal and includes major events such as trainings, purchases, pre/posttests, and activities. The time frame and the amount of time each event will take are included.	The timeline includes most major events such as trainings, purchases, pre/posttests, and activities. The time frame and the amount of time each event will take are included.	The timeline may be missing major events such as trainings, purchases, pre/posttests, and activities. The time frame and the amount of time each event will take are not included.
BUDGET	Items are grouped into categories and include an estimated cost. A strong justification is written for all large-ticket items. The budget does not include items restricted by the granting agency.	Items are grouped into categories and include an estimated cost. An adequate justification is written for all large-ticket items. The budget does not include items restricted by the granting agency.	The budget includes items restricted by the granting agency and/or exceeds the amount granting agency will fund.
EVALUATION METHODS	Several measurable methods are used that demonstrate the impact of the proposed grant on students and community.	At least one measurable methods is used that demonstrate the impact of the proposed grant on students and community.	The evaluation methods are not included and/or do not adequately demonstrate the impact of the proposed grant on students and community.

ITEM	EXEMPLARY	ADEQUATE	NEEDS IMPROVEMENT
SUSTAINABILITY	Strong evidence exists for future sustainability of project.	Limited evidence exists for the sustainability of project.	No evidence exists for the sustainability of project; the project is a onetime event.
RÉSUMÉ OR VITAE	The résumé includes the most recent and significant activities and is limited to two pages. The activities strongly convey the ability to successfully carry out a funded project.	The résumé includes most of the recent and significant activities and is limited to two pages. The activities convey the ability to successfully carry out a funded project.	The résumé is out of date; activities listed do not convey the ability to carry out a funded project.
LETTERS OF SUPPORT	The letters convey enthusiasm and strongly support proposal. The letters are signed, dated, and on letterhead.	The letters convey some enthusiasm and support for the proposal. The letters are signed, dated, and on letterhead.	The letters convey weak enthusiasm and support. The letters may be missing a signature or date or may not be typed on letterhead.
LESSON PLAN	The lesson is strongly tied to the grant proposal and involves authentic inquiry or data collection.	The lesson is tied to the grant proposal and involves data collection.	The lesson is weakly tied to the grant proposal and consists of a worksheet.
GRAMMAR/VOICE	Passion for the idea comes across clearly with few grammatical mistakes, and the proposal is concise and written in third person.	The idea is clearly conveyed but may lack passion and/or may contain some grammatical mistakes that do not interfere with readability.	The overall writing is confusing and/or unclear. There are many grammatical mistakes, which detracts from readability.
SUMMARY OR ABSTRACT	The summary/abstract is concisely written with clear transitions. It heightens interest in learning more about the proposal and includes all the key information (needs statement, target audience, objectives, project description, methods, evaluation, dissemination plan).	The summary/abstract is clearly written for the most part, with all key information included. It does not effectively heighten the readers' curiosity.	The wording is unclear or may not use transitions. It may be missing some key information and may leave the reader confused.

Using the NGSS as a Blueprint

APPENDIX
3

Using the *NGSS* as a Blueprint

1. What is the disciplinary core idea( ) you plan to use as a basis for your grant?

2. List the performance expectation(s) that you will use to develop your objectives.

3. How can you connect the performance expectation(s) to a real-world setting? (*Tip:* Examine the connections to the nature of science and engineering, technology, and applications for science.)

4. In the table below, list objectives/goals that you would like your students to be able to accomplish and their associated science and engineering practices.

OBJECTIVES/GOALS	SCIENCE AND ENGINEERING PRACTICES

5. What crosscutting concepts can be incorporated into your objectives/goals?

6. How can you connect *Common Core State Standards* to your grant? (*Tip:* Examine the *Common Core* Connections found in the *NGSS*.)

Grant Listings and Proposal Resources

APPENDIX
4

Grant Listings

(*Note:* These web addresses are accurate at the time of publication.)

- CDW-G. 2015. *GetEdFunding. www.getedfunding.com/c/index.web?nocache@3+s@T3hBdvXsr5hY2.*

- Edutopia. 2015. *The big list of educational grants and resources. www.edutopia.org/grants-and-resources.*

- Grants for Teachers. 2015. *Upcoming grant deadlines. www.grantsforteachers.net.*

- Grant Wrangler. 2015. *Featured grants for teachers. www.grantwrangler.com.*

- National Science Teachers Association. 2015. *Your elementary classroom. www.nsta.org/elementaryschool.*

- National Science Teachers Association. 2015. *Your middle school classroom. www.nsta.org/middleschool.*

- National Science Teachers Association. 2015. *Your high school classroom. www.nsta.org/highschool.*

- National Science Teachers Association. 2015. *NSTA calendar. www.nsta.org/publications/calendar.*

- Teach. 2015. *Grants for teachers. http://teach.com/what/grants-for-teachers.*

- Teachers Count. 2015. *Grants for teachers. www.teacherscount.org/grants.*

- Teacher Planet. 2015. *Grants 4 teachers. www.grants4teachers.com.*

Grant Proposal Writing Resources

Grant proposal writing workshops are often available at local community colleges and universities or as online classes. Additional resources are listed below.

- Foundation Center. 2015. *http://foundationcenter.org.*

- McCabe, C. 2016. *Write a grant. www.nea.org/home/10476.htm.*

- The Monsanto Fund. 2016. *Best practices in grant writing. www.monsantofund.org/_pdfs/General-Grantwriting-Webinar.pdf.*

APPENDIX 5

FAQs About the Grant Proposal Writing Process

FAQs About the Grant Proposal Writing Process

I need funding for technology for my classroom. Where can I find grants that fund technology?

Although there are some organizations willing to fund technology merely for the sake of providing additional resources for students, it has been our experience that being able to explain to the funder how you will use the technology and listing the anticipated learning outcomes make for a much more powerful request than merely stating that you need technology. You might want to ask yourself the following questions:

- Why do I need the technology?
- What specific actions will my students perform using the technology?
- What will my students learn from their actions?
- Could the same learning occur without the technology?

Is it okay to write a grant that only affects a handful of children?

It depends. In many cases, grant funders are not overly concerned with the actual number of students involved as long as your proposal has merit and fits with their mission. Whether your entire school district benefits from your proposal or if only a small class of AP biology students benefit is not nearly as crucial as what you hope to accomplish. One way to increase the number of participating students is to include a mentoring piece in which your students work with younger students. Another option would be to partner with another teacher either in your school or to connect virtually for the purpose of replicating and comparing student data.

I teach in a well-to-do or private school. Will it be difficult for me to secure grants?

Some grants specify that to qualify your school needs to be a Title I school or specifically meet the needs of underserved populations. Always check the grant requirements to make sure that the grant matches your particular teaching situation. Although it may be more difficult to illustrate need within a wealthier school district, consider taking a look at the trends within your district. Is the number of special needs students or students receiving free and reduced lunch increasing? If so, consider using that information to strengthen your case. Alternately, there are many granting agencies that will provide funds for well-written proposals regardless of the demographics of the school district.

Where can I locate demographic information?

You can begin collecting data about your school or school district by speaking with the person in your district who handles district finances. You can also go to your state's department of education website to find information about your school district. Alternately, you can explore the National Center for Education Statistics (*www.nces.ed.gov*), which is the federal government office in charge of education statistics. Use the "school search" feature to retrieve information related to enrollment by ethnicity, teacher-student ratio, and gender. Using this tool, you can compare statistics between your school and a neighboring school for the purpose of strengthening your argument.

What is the difference between a goal and an objective?

These terms are often used interchangeably. Let the funding organization guide the term you use. If the organization uses the term *goal* instead of *objective,* then mirror their terminology and use the word *goal* in your writing. Other terms that you may run into when writing grants include *outcomes, deliverables,* and *purpose.*

What happens if I don't meet my goals and objectives?

Although you may not meet all of your goals and objectives, it is highly likely that you will be able to provide some evidence of success. In your final grant report, you will have an opportunity to explain what went well and things that could be improved. This is part of the learning process. Although funders really enjoy reading success stories, they realize that things don't always go as planned.

What about standards?

Linking your state standards or national standards to your proposed objectives is an excellent idea. Rather than simply making a list of standards, be sure to clarify the relationship between the standards, your proposal, and student learning; many grant readers are not professional educators.

Is it okay to include methods from a kit or curriculum?

Provided that the methods engage students in inquiry, the use of kits or curriculum is fine; however, we suggest that you consider how the students can apply what they learn from the methods to a real-world setting. Are the students able to manipulate variables or develop their own inquiry or do the methods confine them to a scripted lesson? If you like some curriculum or kit activities, think how they could be modified to allow your students more exploration and inquiry while working on the project.

How much should I charge for in-kind services?

One measurement of in-kind services would be to base the amount on how much a teacher would make via a supplemental contract for writing curriculum or performing some other administrative task. If you are planning on coaching a science team as an in-kind service, you may want to base the amounts on a coach's salary. Regardless of what you are doing, don't forget to document the hours you worked at contributing toward making your project successful.

What happens to the equipment if I leave my position?

In most cases, the check will be written to the school district. Under no circumstances should you take the equipment with you if you leave the district. If you transfer to a new setting in your district, it would be best to approach your principal to ask permission to take the equipment with you to your new classroom (assuming you would have use for it in your new position).

The grant can be awarded to me personally or to the school. What should I do?

Some grants give you a choice of writing the grant directly to you or to your school district. We discourage you from accepting the money directly for a couple of reasons. For one, you will find that taxes may be taken out of the funds prior to you receiving the money. Two, you may have to claim the money as income. Either way, it is less money for your classroom and more bookkeeping in the long run.

What if I propose to present a session at a conference as my dissemination plan and I don't get accepted?

This really isn't a source of concern as there are many outlets for dissemination other than a conference. Consider venues within your school, department, school district, and community if your initial plans do not pan out.

How do you provide a sustainability plan for a project that includes a field trip?

When writing a proposal to fund a field trip, it will be important for you to consider how future field trips will be funded. Will your district help support such trips? Will you ask parents to pay the cost in future years? Will you fundraise? Figuring out how to build in sustainability will make a good idea become part of your annual curriculum.

I need to include a résumé with the grant application. How long should the résumé be?

Many organizations will give you a page length; make sure to adhere to the length. Résumés more than two pages long should be edited to include only the most recent and significant activities. When writing, avoid educational jargon that may be unfamiliar to the reader.

I have problems summarizing my proposal in the required word length set by the funding agency. Do you have any suggestions?

Write a summary that includes everything you think it is important for to convey the project idea. Start to work with that summary and try to reword and shorten the summary until you reach the required number of words. You may want to give the grant proposal to a friend and ask him or her to write the summary (perhaps you know an English teacher or social studies teacher who could assist you). This has the added benefit of your friend being able to identify any unclear elements of your proposal.

My principal told me to write the letter of support and to give it to him or her to sign. What should I do?

We do not recommend that you write your own letter of support simply because it is difficult to change personal writing style. If you write your own letter, you run the risk of anyone reading your grant concluding that you wrote the letter of support. If you find yourself in this situation, we suggest that you ask a colleague to read over the grant and write the letter of support.

I need to submit a sample lesson plan with the grant application. How long does the lesson have to be?

Typically, funders who ask for a lesson do not provide much detail regarding what they would like to see. Using your proposal's objectives, identify a lesson that you feel would best show your ability to plan, carry out, and assess for student learning. This is an area where you have a lot of expertise, and it will be a relatively easy task for you to accomplish. It is also a great vehicle for you to demonstrate your passion for your project.

What if there appears to be a problem with the online submission?

Call or e-mail the grant manager and explain your concern. They are generally very happy to assist because they want to receive proposals.

Grant Resources at State Departments of Education

APPENDIX
6

STATE	DEPARTMENT OF EDUCATION GRANT MANAGEMENT OFFICE
ALABAMA	Alabama Department of Education Office of Financial Management *www.alsde.edu*
ALASKA	Alaska Department of Education Office of Project Management and Permitting *education.alaska.gov*
ARIZONA	Arizona Department of Education Grants Management *www.azed.gov/grants-management*
ARKANSAS	Arkansas Department of Education *www.arkansased.gov/divisions/communications/grants*
CALIFORNIA	California Department of Education Finance & Grants *www.cde.ca.gov/fg*
COLORADO	Colorado Department of Education *www.cde.state.co.us/cdefisgrant*
CONNECTICUT	Connecticut Department of Education Bureau of Grants Management *www.sde.ct.gov/sde/cwp/view.asp?a=2680&Q=320640*
DELAWARE	Delaware Department of Education Finance and Operations Office *www.doe.k12.de.us/site/default.aspx?PageID=1*
FLORIDA	Florida Department of Education Contracts, Grants & Procurement Office of Grants Management *fldoe.org/finance/contracts-grants-procurement*
GEORGIA	Georgia Department of Education Science, Technology, Engineering, and Mathematics (STEM) *www.gadoe.org/Curriculum-Instruction-and-Assessment/CTAE/Pages/STEM.aspx*
HAWAII	Hawaii State Department of Education *www.hawaiipublicschools.org/Pages/Home.aspx*
IDAHO	Idaho State Department of Education *http://sde.idaho.gov*

STATE	DEPARTMENT OF EDUCATION GRANT MANAGEMENT OFFICE
ILLINOIS	Illinois State Board of Education Innovation and Improvement The Grant Administration *www.isbe.net/grants/default.htm*
INDIANA	Indiana Department of Education Grants Management *www.doe.in.gov/grantsmgt*
IOWA	Iowa Department of Education *www.educateiowa.gov*
KANSAS	Kansas State Department of Education *www.ksde.org/Default.aspx?tabid=532*
KENTUCKY	Kentucky Department of Education Office of Administration and Support Division of Budget and Financial Management *education.ky.gov/districts/fin/Pages/Grant%20Information.aspx*
LOUISIANA	Louisiana Department of Education Grants Management *www.louisianabelieves.com/funding/grants-management*
MAINE	Maine Department of Education *www.maine.gov/doe*
MARYLAND	Maryland State Department of Education *www.marylandpublicschools.org*
MASSACHUSETTS	Massachusetts Department of Education Current Grant Funding Opportunities *www.doe.mass.edu/Grants*
MICHIGAN	Michigan Department of Education Grants *www.michigan.gov/mde/0,4615,7-140-5236---,00.html*
MINNESOTA	Minnesota Department of Education Grants *http://education.state.mn.us/MDE/SchSup/Grant/index.html*

STATE	DEPARTMENT OF EDUCATION GRANT MANAGEMENT OFFICE
MISSISSIPPI	Mississippi Department of Education Healthy Schools Grant Opportunities *www.mde.k12.ms.us/OHS/FO*
MISSOURI	Missouri Department of Elementary and Secondary Education Grants & Resources *http://dese.mo.gov/grants-resources*
MONTANA	Montana Office of Public Instruction Finance and Grants *opi.mt.gov/Finance&Grants/Index.html*
NEBRASKA	Nebraska Department of Education Grants Management System *www.education.ne.gov/GMS2*
NEVADA	State of Nevada Department of Education Grants *www.doe.nv.gov/About/Grants*
NEW HAMPSHIRE	New Hampshire Department of Education Online Grants Management System Handbook *www.education.nh.gov/data/documents/grants_manage_handbook.pdf*
NEW JERSEY	State of New Jersey Department of Education Office of Grants Management *www.state.nj.us/education/grants*
NEW MEXICO	New Mexico Public Education Department Fiscal Grants Management *ped.state.nm.us/ped/FiscalGrantsMgmntIndex.html*
NEW YORK	New York State Education Department *www.nysed.gov/finance-business/funding-grants*
NORTH CAROLINA	North Carolina Department of Public Education Finance & Grants *ec.ncpublicschools.gov/finance-grants*
NORTH DAKOTA	North Dakota Department of Public Instruction Grants *www.dpi.state.nd.us/grants/index.shtm*

STATE	DEPARTMENT OF EDUCATION GRANT MANAGEMENT OFFICE
OHIO	Ohio Department of Education Grants *education.ohio.gov/Topics/Finance-and-Funding/Grants*
OKLAHOMA	Oklahoma State Department of Education Grants and Nominations *ok.gov/sde/grants-and-opportunities*
OREGON	Oregon Department of Education Grants E-Grant Management System (EGMS) *www.ode.state.or.us/search/results/?id=69*
PENNSYLVANIA	Pennsylvania Department of Education *www.education.pa.gov*
RHODE ISLAND	Rhode Island Department of Education Funding Sources *www.ride.ri.gov/FundingFinance/FundingSources.aspx*
SOUTH CAROLINA	South Carolina State Department of Education Grants Program *http://ed.sc.gov/about/division-for-legal-affairs/grants*
SOUTH DAKOTA	South Dakota Department of Education Grants Management System *http://ed.sc.gov/finance/grants*
TENNESSEE	Tennessee Department of Education Finance and Monitoring *www.tn.gov/education/section/finance-and-monitoring*
TEXAS	Texas Education Agency Finance and Grants *tea.texas.gov/Finance_and_Grants/Grants*
UTAH	Utah State Office of Education Academic Service Learning—Grant Management *www.schools.utah.gov/adulted/Directors---Coordinators/Grants.aspx*
VERMONT	Vermont Department of Education Small Schools Support Grants *education.vermont.gov/data/small-schools-support-grants*

STATE	DEPARTMENT OF EDUCATION GRANT MANAGEMENT OFFICE
VIRGINIA	Virginia Department of Education Budget and Grants Management *www.doe.virginia.gov/school_finance/budget/index.shtml*
WASHINGTON	State of Washington Office of Superintendent of Public Instruction Finance & iGrants *www.k12.wa.us/Finance/default.aspx*
WEST VIRGINIA	West Virginia Department of Education Grant Opportunities *wvde.state.wv.us/healthyschools/Grants.html*
WISCONSIN	Wisconsin Department of Public Instruction Grants and Financial Assistance *dpi.wi.gov/grants*
WYOMING	Wyoming Department of Education Grants *edu.wyoming.gov/beyond-the-classroom/grants*

DISTRICT OR TERRITORY	DEPARTMENT OF EDUCATION GRANT MANAGEMENT OFFICE
AMERICAN SAMOA	American Samoa Department of Education *www.doe.as*
COMMONWEALTH OF THE NORTHERN MARIANA ISLANDS	Commonwealth of the Northern Mariana Islands Public Schools *www.cnmipss.org*
DISTRICT OF COLUMBIA	District of Columbia Public Schools Fellowships, Grants, and Awards for Teachers *dcps.dc.gov/page/fellowships-grants-and-awards-teachers*
GUAM	Guam Department of Education *sites.google.com/a/gdoe.net/gdoe*
PUERTO RICO	Puerto Rico Department of Education *www.de.gobierno.pr*
U.S. VIRGIN ISLANDS	Virgin Islands Department of Education Federal Grants and Audits *http://vide.vi*

National Science Teachers Association Position Statements

APPENDIX
7

National Science Teachers Association (NSTA) Position Statements

Note: All position statements are available on the NSTA website at *www.nsta.org/about/positions.*

Accountability

Aerospace Education

Animals: Responsible Use of Live Animals and Dissection in the Science Classroom

Assessment

Beyond 2000—Teachers of Science Speak Out

Competitions, Science

Computers in Science Education, The Use of

Disabilities, Students with

Early Childhood Science Education

E-Learning in Science Education, The Role of

Elementary School Science

English Language Learners, Science for

Environmental Education

Evolution, The Teaching of

Gender Equity in Science Education

High School Science, Learning Conditions for

Induction Programs for the Support and Development of Beginning Teachers of Science

Informal Environments, Learning Science in

Inquiry, Scientific

International Science Education and the National Science Teachers Association

K–16 Coordination

Laboratory Investigations in Science Instruction, The Integral Role of

Leadership in Science Education

Liability of Science Educators for Laboratory Safety

Metric System, Use of the

Middle Level Students, Science Education for

Multicultural Science Education

Nature of Science

The *Next Generation Science Standards*

Parent Involvement in Science Learning

Preparation, Science Teacher

Professional Development in Science Education

Professionalism for Science Educators, Principles of

Quality Science Education and 21st-Century Skills

Research on Science Teaching and Learning, The Role of

Safety and School Science Instruction

Societal and Personal Issues, Teaching Science and Technology in the Context of

Teaching Awards

APPENDIX
8

Awards Sponsored by the National Science Teachers Association (NSTA)

NSTA Distinguished Service to Science Education Awards

NSTA Distinguished Teaching Awards

DuPont Pioneer Excellence in Agricultural Science Education Award

Maitland P. Simmons Memorial Award for New Teachers

Northrop Grumman Foundation Excellence in Engineering Education Award

NSTA Fellow Award

PASCO STEM Educator Awards

Robert E. Yager Foundation Excellence in Teaching Award

Robert H. Carleton Award

SeaWorld Parks and Entertainment Outstanding Environmental Educator of the Year

Shell Science Teaching Award

Shell Urban Science Educators Development Award

Sylvia Shugrue Award for Elementary School Teachers

Vernier/NSTA Technology Awards

Dr. Wendell G. Mohling Outstanding Aerospace Educator Award

Other Awards

American Association of Petroleum Geologists

Excellence in Teaching Award

American Association of Physics Teachers

Paul W. Zitzewitz Award for Excellence in K–12 Physics Teaching

American Chemical Society

James Bryant Conant Award in High School Chemistry Teaching

Outstanding High School Chemistry Teacher of the Year

Paul Shin Memorial Outstanding High School Chemistry Teacher Award

American Geosciences Institute

Edward C. Roy Jr. Award For Excellence in K–8 Earth Science Teaching

ASM International Materials Information Society

Kishor M. Kulkarni Distinguished High School Teacher Award

Bradley Stoughton Award for Young Teachers
Albert Easton White Distinguished Teacher Award

Council for Elementary Science International
March C. McCurdy International Award
CESI Muriel Green Award for New Teachers

International Society for Technology in Education
ISTE Outstanding Young Educator Award
Kay L. Bitter Vision Award

National Association of Biology Teachers
Outstanding New Biology Teacher Achievement Award
Biology Educator Leadership Scholarship (BELS)
Ecology/Environmental Science Teaching Award
Evolution Education Award
Genetics Education Award
The Kim Foglia AP Biology Service Award
Outstanding Biology Teacher Award
Outstanding New Biology Teacher Achievement Award
The Ron Mardigian Biotechnology Teaching Award

National Earth Science Teachers Association
Outstanding Earth Science Teacher

National Middle Level Science Teachers Association
Paul deHart Hurd Award for Outstanding Teaching

INDEX

*Page numbers printed in **boldface** type refer to tables or figures.*

Be a WINNER!

Index

Index

Index

Index